Fɪɢ. 1. Fredrik Adam Smitt. Portrait from Anonymous (1905).

TRANSACTIONS

OF THE

AMERICAN PHILOSOPHICAL SOCIETY

HELD AT PHILADELPHIA
FOR PROMOTING USEFUL KNOWLEDGE

———

NEW SERIES—VOLUME 63, PART 7
1973

———

F. A. SMITT, MARINE BRYOZOA, AND THE INTRODUCTION OF DARWIN INTO SWEDEN

THOMAS J. M. SCHOPF

Department of Geophysical Sciences, University of Chicago

AND

EDWARD L BASSETT

Department of Classics, University of Chicago

———

THE AMERICAN PHILOSOPHICAL SOCIETY
INDEPENDENCE SQUARE
PHILADELPHIA

December, 1973

Library of Congress Catalog
Card Number 73-86617
International Standard Book Number 0-87169-636-3

F. A. SMITT, MARINE BRYOZOA, AND AN APPLICATION OF THE DARWINIAN PRINCIPLE OF DESCENT IN 1868

Thomas J. M. Schopf and Edward L. Bassett

CONTENTS

INTRODUCTION

It can safely be stated that most present zoological taxonomists have little useful knowledge of Latin. Ironically, "general articles" which were written one hundred or more years ago in Latin just so that they would be read are now unintelligible. One casualty of this trend has been F. A. Smitt's classic article "Bryozoa Marina" (1868) on bryozoan evolution and the principles of classification. Smitt was the first Swedish naturalist to accept and apply Darwin's theory of evolution.

Often in the one hundred years since Smitt's article, taxonomists have been indicted and denigrated for keeping their guiding principles under cover, perhaps revealing them only to the few who know the secret language of the taxonomy of that group. In Smitt, it is more a question of knowing the secret language of Latin, for we have a practicing taxonomist and theore-

tician who explained why he used the taxonomic arrangements that he did. Secondly, he was an extremely strong proponent of the biogenetic law, namely that there is a fundamental relationship between the sequence of development of an individual organism and the evolutionary history of its species ("ontogeny recapitulates phylogeny").

Thirdly, Smitt is the founder of modern views of classification of living Bryozoa, a group with more than 3,000 extant species. As his contemporary, the renowned naturalist Rev. Thomas Hincks, B.A., F.R.S., wrote:

> To estimate rightly the work which the Swedish naturalist has accomplished in this department of zoology we must remember that it is not a mere revision of an existing system that we owe to him, but the institution of a new system, resting on new foundations, and implying a new interpretation of the facts with which it deals. His distinctive merit is that he substituted zooecial for colonial characters as the proper basis of a natural arrangement, thus giving a new direction to research and preparing the way for a system which should rest not on mere superficial resemblances, but on genetic affinity. (Hincks, 1890: pp. 83–84.)

Finally, as initially developed by Smitt in another paper (Smitt, 1865b), but discussed in the paper translated here, he elucidated the mechanism of growth in the Bryozoa. These ideas on growth are being hotly debated in the literature a century later.

For all of these reasons, Smitt's views deserve much wider recognition and attention than generally is the case. Partly to satisfy our own curiosity, and partly to make this paper available to others, and thus replace it in its rightful position in the world literature, we prepared this translation and commentary.

The translation was begun in the spring of 1969 by Mr. Donald W. Layton, Jr., then a senior in classics at Lehigh University, and by Schopf. At that time, assistance in some difficult passages was given by Professor J. A. Maurer, Department of Classics, Lehigh University. When Schopf came to the University of Chicago in the fall of 1969, he took the notes and text to Professor E. L. Bassett of Classics for comment. In the course of time, Bassett and Schopf together reworked the entire translation in detail, and that version is presented here. It was completed in the spring of 1971.

In order to place the translation in context, we include sections on the significance and importance of Smitt (chiefly by Schopf) and the Latinity of Smitt (chiefly by Bassett).

Smitt was a well-known but not a famous scientist. His most complete biography appears in the Royal

Swedish Academy (anonymous, 1905), but shorter summaries of his life have appeared in standard biographical sources (e.g., *Svenskt Biografiskt Handlexikon*, 1876; *Svenskt Biografiskt Lexikon*, 1907; *Nordisk Familjebok*, 1917, with portrait; *Svensk Uppsalagsbok*, 1957).

All quotations rendered in English from Smitt's work are by us, except for the 1872 article which was published in English. Original letters of Smitt are given by Danielsson (1963–1964: pp. 192–194; 1965–1966: p. 273) in an extremely useful article on the introduction of Darwinism into Sweden. We considered republishing the Latin text with our translation but the journal in which it was presented is not excessively rare and should be available to interested readers.

We are grateful for advice on the translation of passages from Swedish, or for commenting on aspects of the manuscript, from several persons, particularly Dr. Tore Wretö, Professor Sten Lindroth, Mrs. Ida Thompson, Professor Stig M. Bergström, Mrs. Öse Manheim, and Dr. Frank T. Manheim.

F. A. SMITT

LIFE

SUMMARY OF LIFE

Fredrik Adam Smitt was born on 9 May, 1839, in the Swedish port of Halmstad. His father, Dr. Johan Jakob Smitt, was the provincial physician from 1838 to 1853. Dr. J. J. Smitt married Anna Carolina Maria Strömberg, who was the daughter of the Danish vice consul in Strömstad, H. J. Strömberg. Johan and Anna Smitt had eight children of whom Fredrik was the third.

When Fredrik was fourteen, his father died, but Fredrik was able to pursue his studies supported by a scholarship from his older brother Ertman J. Smitt and from C. Hammar, a merchant and member of parliament. In the fall of 1855 at the age of sixteen, Fredrik entered Lund University. He transferred to Uppsala University a year later. This change also corresponded with the retirement of Sven Nilsson from the professorship in zoology at Lund, and the appointment of Wilh. Lilljeborg to the same position at Uppsala. Lilljeborg directed Smitt's attention to the aquatic arthropod *Daphnia* on which Smitt published his first paper (1861). Of greater influence was Sven Lovén at the Naturhistoriska Riksmuseum in Stockholm, who inspired Smitt to take up the marine group Bryozoa (Danielsson, 1963–1964: p. 194). Smitt was graduated from Uppsala in 1859, and became the "amanuens" in charge of the Marklin zoological collections at Uppsala University.

Two years later in 1861, Fredrik took part in the Swedish expedition to Spitsbergen where he shared the position of naturalist with Goes on the ship *Magdalena*. Smitt became a candidate for the doctor of philosophy examination at Uppsala on his return in 1862. His thesis was "Contributions to the Knowledge of the Development of Marine Bryozoa." The following year (1863) he received his degree and became a docent ("associate professor") in zoology. From 1865 to 1867, Fredrik traveled on the continent as a Byzantine Fellow (supported by a travel-fund established in 1808 by P. O. Asp (1745–1808), who had been ambassador to Constantinople in 1790). Smitt spent the greater part of his time on zoological studies in Paris and Copenhagen. Upon returning to Sweden in 1868, he participated in a new expedition with A. E. Nordenskiöld to Spitsbergen and Bear Island. Smitt is said to have won the admiration of his comrades for his energy, strength, and toughness which he displayed by sorting dredge hauls from ice cold water hours at a time. The following year he accompanied the frigate *Joséphine* as the zoologist on her expedition to Portugal, the Azores, England, and North America.

In 1871 F. A. Smitt was appointed the successor to Carl Jakob Sundevall as professor for the Department of Vertebrates at the Naturhistoriska Riksmuseum in Stockholm. He was employed there until his death on 19 February, 1904. These curatorial interests were supplemented by an appointment for teaching zoology in a Stockholm high school in 1879. From 1892 to 1904 he was the principal administrator of the museum. In 1874 he married Freja Brigitta Karolina Palman; they had two sons and a daughter.

LANGUAGE TRAINING

F. A. Smitt's training in languages was probably very extensive. In Western Europe, at least through the time of Linnaeus (1707–1778, *Systema Naturae*, 10th edition, 1759), Latin was the language of science. Following the French Revolution, French, English, and German gradually superseded Latin in chemistry and physics. In 1825 in Sweden, a committee was designated to review the whole schooling system. It recommended that there be two "lines of study." In one of these, classical languages would continue to be used; in the other, sciences would be emphasized, especially chemistry; English was recommended for both lines of study. Although some schools perhaps adopted the recommendation, this program was not instituted throughout the country until a decree of 1849. Since he was born in 1839, it appears as though F. A. Smitt would have begun his studies in a "classical line," and to have been switched to the "modern science line" later on.

Smitt's published work includes approximately 3,400 pages of text, and is written in five languages. Approximately 86 per cent of the literature is in Swedish, 5.7 per cent English, 5.7 per cent French, 2.5 per cent Latin, and 0.1 per cent German. In addition, he translated a three-volume 1,900-page German book on general animal biology into Swedish.

It is not surprising that Smitt wrote a scientific article in Latin because of the tradition of using Latin in taxonomic biologic work. Other works in Latin in

Sweden in the 1800's include C. A. Agardh's papers on algae (1820's), Elias Fries's *Systema Mycologicum* (1821–1832), and works on trilobites by Angelin (1851–1878). To this day a Latin diagnosis must accompany the description of a newly discovered plant species (but not an animal), although the proposal also to include animals is sometimes made (Annoscia, 1968). Additionally, in Smitt's time it may well have been easier for his English and German colleagues, some of whom were trained in the church (e.g., Canon Norman, Rev. T. H. Hincks), to read Latin than Swedish, "a language available to but few naturalists" as Alexander Agassiz once remarked. The combination of these factors provides a plausible argument as to Smitt's reasons for writing in Latin.

SCIENTIFIC WORK

Smitt's scientific work is dominated by the theme of determining evolutionary relationships between groups of organisms. Smitt was the first Swedish scientist to write a major scientific paper explicitly using the Darwinian principle of descent (Smitt, 1865a; see Danielsson, 1963–1964: p. 192). In his address to the Third International Congress of Zoology in 1895, he wrote "For a very long time—that is, more than 30 years— I have sought in my particular studies a general method for finding the affiliation of animal species" (Smitt, 1896a: p. 235). This guiding question comes up again and again in his university and early post-university career (approximately 1860–1878) which was dedicated to the study of colonial invertebrates, the Bryozoa, and in his later career during which he concentrated on vertebrates (fish). The general approaches that he used to investigate the question of animal relationships were novel for his day, and remain important for the present day. The question he was asking is: How does one infer evolutionary relationships from morphology?

RESEARCH ON BRYOZOA

Smitt's earliest work on Bryozoa concentrated on the process of growth. The earliest formed individuals are often simpler and appear to be more primitive than later formed individuals within the same bryozoan colony. Smitt tried to relate this fact to the growth process for the colony as a whole.

Before going further, a few words on the nature of Bryozoa are in order, using the terminology of the translation. In Smitt's sense, this was a class (now Phylum) that included two major groups: the ectoprocts and the entoprocts. Nearly all of the translated article concerns only one of these groups, the ectoprocts, and further remarks apply only to them. These colonial animals are common but small and obscure (*ca.* 1–5 cm. for many colonies), thus are often overlooked. Bryozoa are phylogenetically important since they and the closely related brachiopods and phoronids share characteristics with the annelid-arthropod line of evolu-

tion on the one hand, and the echinoderm-chordate line on the other hand. Bryozoa are taxonomically very difficult and they have languished in the corners of both systematic zoology and animal physiology for decades. Hence it is remarkable, in a sense, that Smitt did so much with so little.

A feeding crown of tentacles extends through the orifice of either a tubular or a boxlike exoskeleton. In some groups, when the tentacles are withdrawn through the orifice, a lid or operculum closes the orifice. In boxlike individuals, the frontal surface (which includes the orifice) may either be calcified or membranous; if membranous, the frontal surface is then said to have an opesium or region of the opesium. The skeleton may have considerable sculpturing, papilli, and spines. The colony may have individuals modified for attachment, called tendrils or rootlets, or it may have individuals modified for defense known as avicularia. When the modified operculum (called a mandible) of the avicularium closes, the avicularium, as Charles Darwin wrote, can grab a needle "so firmly the branch might be shaken" (Darwin, 1839: p. 259). Sometimes the defense organs are in the form of elongate, whiplike, flexible rods called vibracularia.

In addition to growing by asexual budding to form a colony, bryozoans have a sexual generation. The larvae may be brooded in structures called ovicells or ooecia (singular "ooecium"), literally egg-chambers. After existing in an ooecium for a few days, a larva is released to the seawater where for most species it is passively carried for a few hours, and then settles and attaches to a solid substratum. A newly attached individual sometimes has a morphology entirely unlike the individuals which are then budded from it; if so, it is often referred to as a *tata*. Each settled individual buds new individuals asexually, forming a colony again.

At the present time, more than 3,000 distinct species of these animals are living. In the fossil record is evidence for another 15,000 species, the oldest of which have been reported from the Cambrian. Nearly all living and extinct species are marine, and they occur from the Antarctic to the Arctic, from the intertidal to at least 7,500 meters depth.

COMMON-BUD CONCEPT

Smitt's inaugural dissertation emphasized the morphology and mode of formation of colonies. It resulted in a paper on bryozoan development and growth which won in 1865 the Florman Award of the Swedish Academy of Science, given for excellence in studies of physiology or anatomy.

This paper (Smitt, 1865b) developed the notion of the "common bud," which is an abbreviated way of saying that there is an all-colony growth zone due to the sharing of the coelomic cavity by the whole colony. This sharing results in a coordinated colonial control over the colony growth form. During the past century, this concept has been the key factor in understanding growth in tubular Bryozoa, one of the two major groups of these animals (the other group, the boxlike Bryozoa, are dealt with below). Borg in 1926 elaborated on the common-bud concept in his academical dissertation, and his work is now the fundamental reference for modern studies on the growth of tubular Bryozoa. (Perhaps part of the reason for this is that Borg wrote in English whereas Smitt used Swedish.) Very recent reviews of the growth of tubular Bryozoa have emphasized the common-bud concept as critical to understanding their growth (Boardman and Cheetham, 1969; Ryland, 1970).

Acceptance of the common-bud concept for boxlike Bryozoa has not been nearly as rapid or universal. Borg (1926), for example, stated that it did not apply to the major groups of these forms now in existence, namely the Cheilostomata. Elias (1971) summarizes reasons for doubting the existence of an all-colony growth zone in cheilostomes. However, only in the past three or four years have publications appeared which include histochemical data on mineralized skeletons, and which were obtained from the point of view of constructing models of growth in cheilostomes in general (Banta, 1968, 1969, 1970; Tavener-Smith and Williams, 1970; Schopf and Travis, 1968). These latest papers confirmed the existence of a coelomic extension in the position where it might be able to mediate growth over the colony as a whole. Resolution of this question is certain to be a major issue of those studying Bryozoa in the next few years.

The application of the common-bud concept in understanding the growth of cheilostomes is thus apparently close to the position advocated by Smitt more than one hundred years ago. In his only statement on this topic in English, he wrote,

It is principally from the Cyclostomata that this theory most easily will be understood; but even in the Chilostomata, although there the individual life is more developed, the commonness of the budding is perceptible, though it gradually approaches the simple budding by the uniserial forms (*Eucratea*, etc) (Smitt, 1872*b*: p. 247).

CLASSIFICATION OF BRYOZOA AND ACCEPTANCE
OF EVOLUTION

It is impossible to estimate too highly the thoroughness of research on which Prof. Smitt's classification rests, and the important contribution which he has made towards a natural system, however much we may be disposed to dissent from some of his results (T. Hincks, 1880: p cxiii).

Smitt followed his developmental studies with a series of carefully written and beautifully illustrated papers on the marine Bryozoa of Scandinavia. These papers, chiefly published from 1864 to 1868, are the core of work for all subsequent taxonomic investigations in this part of the world. They include more than 100 species described on some 500 pages of text and 24 **plates.**

What Smitt did in bryozoan taxonomy was to downgrade the importance of characters typical of colonies, such as the growth form, or shape, and to emphasize characters of individual zooids. Smitt's conception of bryozoan classification was based on the repetitive modules of the colony, the individual zooids. Hincks (1880) called this a revolutionary step, and indeed it is essentially the modern view.

The papers on taxonomy of Scandinavian Bryozoa were followed during the ensuing decade with taxonomic reports from other places in the world. Of greatest interest is the 1872–1873 two-paper series on bryozoans dredged by Harvard's L. F. de Pourtalès from off the southeastern United States. The approximately 80 species described represent at least a quarter of the currently recognized tropical Atlantic fauna of the Americas. Smitt's careful illustrations make these papers of especially high quality.

Smitt introduced the now widespread term "zooecium" for the individual exoskeletal chamber and accompanying dermal tissues (1865*a*: p. 115). Prior to that time, these chambers of individuals of the colony had been called "cells," which, as Ryland (1967) states, was clearly inappropriate. Smitt's nomenclature for these units is today used in exactly his sense.

The bryozoan literature of the years following the 1859 publication by Darwin of *The Origin of Species* differed little from those before as far as most students of bryozoan taxonomy were concerned. Published in 1868, "Bryozoa Marina" of Smitt is a clear and sharp exception. The importance of recognizing evolutionary lineages in taxonomic work is repeatedly emphasized by Smitt as is the subsequent establishment of a classification derived from these evolutionary discoveries. One test for an evolutionary classification is to predict the future discovery of a fossil or recent species not yet known, and this test is explicitly indicated. Thus this article, on the one hand, is Smitt's philosophical guidebook for any group, and, on the other, is a practical testimony of the evolution of the Bryozoa. In its own way, "Bryozoa Marina" is as much a test of an evolutionary classification as was Darwin's monograph on the barnacles (Ghiselin, 1969; Ghiselin and Jaffe, 1973).

Smitt did not begin to read Darwin's *Origin of Species* until September 18, 1867 (letter to S. Lovén). However, Lovén, who was Smitt's scientific mentor, was an ardent proponent of the theory of descent. In a letter to Lovén, on July 13, 1867, Smitt wrote that "it shall be nice to hear if uncle [an affectionate term for Lovén] will not agree that I have succeeded in proving that you are right: 'we are now at the concept of the changeability of species.'" Later that same summer, Smitt again wrote to Lovén who was planning a trip to England which was to include a visit to Darwin:

When uncle will meet Darwin, it might be of interest if uncle pointed out the circumstances of development which uncle knows to be valid for Bryozoa; these circumstances no doubt offer the best starting point for theorizing the laws of change. Since I have been dependent upon uncle's collections and observations, I have made uncle to some degree an accomplice in my opinions; but uncle can safely try to persuade them to read my works: I believe that they rather will accuse us of carefulness than of too hasty conclusions.

Lovén did not meet Darwin on that trip, and therefore did not tell him of Smitt's work. By the sixth edition of the *Origin of Species* (1872) however, Darwin was referring to Smitt by name as a naturalist who had carefully studied the Bryozoa. (This paragraph is based on the discussion written in Swedish by Danielsson, 1963–1964: pp. 192–194.)

In the course of his taxonomic work, Smitt thoroughly revised the general principles upon which the classification of the Bryozoa might rest. So revolutionary was his approach that the great English naturalist, and highly esteemed student of the Bryozoa, Thomas Hincks, wrote:

We owe to Prof. Smitt the first serious attempt to substitute a natural system for the purely artificial arrangements hitherto in use; and it may at least be said that, if he has not overcome all the difficulties incident to the work and has left many problems unsolved, he has given us the most fruitful suggestions, and may perhaps have struck the track along which future advances must be made (Hincks, 1880: p. cxx).

In order to differentiate and classify species, Smitt's method was first to understand the evolution of the group involved.

The basic question of classification depends upon correctly evaluating the nature of *species*. Often species vary and are joined together by variations in such a way that the distinguishing features are scarcely to be found, and the limits of even the broader divisions may become uncertain. Concerning the Bryozoa, this uncertainty has created great confusion. In order that this situation may be resolved, we shall try to demonstrate that the differences of these species are due to their evolution and that they can be explained for the most part by that principle. (p. 459L, 15E)[1]

The following examples illustrate his working methodology. In his analysis of tubular (cyclostome) Bryozoa, he wrote:

We therefore conclude that the forms of this genus[*Crisia*] should be arranged according to the law of evolution in a sequence whose members, derived from a common origin, keep their own degree of evolution in such a way that each stage [in the sequence] seems to be a fixed species. The result of this is that we can propose as many species as we see more or less fixed stages.
This is also valid for the families which compose the [Cyclostome] suborder Incrustatina. But the more highly evolved they are, the more difficult it is to show a connection among the forms. This is because they have lived through long geological periods during which several intermediary forms have become extinct, and because the

---

[1] A page reference with L means the page in Smitt's Latin text of "Bryozoa Marina"; one with E, the page in our translation.

mode of origin is often obscured by the more intricate organization of the colony. Even though a large number of species have been described from this suborder, I wish to point out that almost no one has made observations about their evolution. (p. 461L, 17E)

Similarly, in describing boxlike cheilostome Bryozoa, he emphasized that:

We wish first to observe that there are almost no traits so fixed that they do not vary in the evolution of forms. Because of this we shall try in vain to describe with definite traits the limits of orders, families, and genera (often also of species). These series, having evolved from the same origin, resemble each other in many ways. Often the forms of individual series are so similar that it is very difficult to decide where a given form is to be referred. Therefore, we are not able to do anything other than follow through the evolution of each form so that we may have a definite decision about its nature. (p. 469L, 21E)

Thus far we have concentrated on the fact that Smitt was a strong proponent of Darwin's idea of descent. However, the specific mechanism driving evolution which was advocated by Darwin, namely natural selection, seemed to Smitt in general to be too weak. After granting that "Natural selection can explain a good deal, and so far as I can see it is a fact that it influences most of the changes of a species," Smitt asked rhetorically in a letter to Lovén (October 4, 1867), "but can it explain them all, and all together?" Later in his general book of 1876 on the development of higher animals, he wrote "natural selection would be inconceivable without its depending upon a selecting person." He refers to experiments on fish that can change color depending upon the background they are on. When the third nerve of these fish is cut, they lose the capability of changing color. Thus, concludes Smitt, the ability to change color is mediated by the nervous system and must be a willful response that became a habit and later on became hereditary. Thus development depended partly on the modifying influence of the surrounding world (which could be called natural selection) but also partly and most characteristically on the innate developmental powers of forms. This teleological idea of development put Smitt squarely in the forefront of those who later came to be called neo-Lamarckians after that famous French scientist's ideas of acquired characters. (This paragraph is based on the discussion written in Swedish by Danielsson, 1963–1964: p. 194; 1965–1966: p. 273.)

Finally, the rapidity with which Smitt accepted the notion of Darwinian descent perhaps accounts for one very peculiar and difficult aspect of his practical taxonomy. In his monograph on Scandinavian Bryozoa, Smitt often arranged the lower taxa into species, forms, and sections, rather like species, sub-species, and sub-sub-species. It is almost as though Smitt were trying to derive a certain species from another *recent* species, and that one from yet another species still living, etc., with many particular transitional forms indicated in the taxonomic arrangement (we are grateful to Dr. Lars

Silén for discussions on this point). If this was indeed Smitt's *modus operandi*, then his practical taxonomy is understandable only when we consider the philosophical basis from which it was derived.

VARIATION WITHIN A COLONY AND THE BIOGENETIC LAW

The origin of variation within a colony was of special interest to Smitt. He emphasized that the genetic potential throughout a single bryozoan colony was the same:

However, if we examine the animals which live united in colonies, whose genealogy of individuals is therefore not to be disputed, we shall be able to investigate more easily the laws and reasons for variation. In this respect the Bryozoa will be of the highest assistance because great variations occur in their colonies. (p. 459L, 15E)

The main evolutionary reason for studying the variation within and between colonies, however, was that this might be a very practical method of determining evolutionary relationships. It would work as follows. If the gradual developmental series of changes within a colony was recapitulating the evolutionary history of that species, then all one had to do was to assemble the various species in a sequence according to the stage of their last formed individuals and an evolutionary classification would result. Smitt wrote:

A species which has a colonial mode of life is composed of individuals that have the same external differences as those which distinguish genera or families in classifications (Smitt, 1896a: p. 236).

Or, in another passage:

We can follow the variations when they are arranged together in series, from the simplest stages up to the most highly evolved forms. This is just as valid for individuals of species and for individuals of colonies, as it is for distinct species. Therefore, the differentiation of a colony shows a series of variations that progresses in the same way that a series of differences progresses among species. (p. 459L, 15E)

Biologists will recognize Smitt's advocacy of the "Biogenetic Law," that ontogeny recapitulates phylogeny. Originally this relationship was proposed because of the general parallel sequence in the embryonic development of higher animals with the hypothesized evolutionary sequence for major divisions of the animal kingdom. The stages of development of higher animals corresponded to "the grades of increasing structural complexity into which the lower animals could be arranged" (Hyman, 1940: p. 273). The degree to which developmental stages "recreate" ancestral adult stages as opposed to merely ancestral embryonic stages was a subject of very great dispute during the nineteenth and early twentieth centuries (DeBeer, 1962).

Smitt evidently intended to pursue to the absolute limit the theme that developmental stages within a colony represented a summation of the evolutionary history of the species. In his "science for the people" lecture of 1898, titled "Darwinism in recent times" (Smitt, 1898b), Smitt wrote:

[In the Bryozoa] the question has reference not merely to organs or parts of individuals, or disappearing stages of development, but to completely developed type examples which reproduced themselves in colonies, and appeared in a more highly developed form, according to their relative position within the colony, until in overdeveloped colonies, the last individuals frequently assumed a simplified form. This provides a microcosm of the general evolutionary law: progress, culmination and regression, which in this case could not be deduced without the use of hypotheses.

Although Smitt and Ernst Haeckel (1834–1919), the leading proponent of the biogenetic law, were contemporaries, the influence of Haeckel on Smitt has not yet been examined. Indeed the influence of Haeckel on Swedish biological theory has not yet been examined (letter from Professor Sten Lindroth to T. J. M. Schopf, April 16, 1971).

The important point for us is that in Bryozoa, the larva is of a very generalized form and that the embryological sequence is greatly confused by a total reorganization of the larva in forming the adult. In contrast, the morphologic sequence of successively budded adult individuals in a colony does show to some extent a progressive change which is not greatly at variance with proposed evolutionary stages as judged from other evidence (Boardman and Cheetham, 1969: p. 229). Smitt supported the biogenetic law in an extreme form since it is in the adult rather than larval relationships that there is a suggestion of its truth. A more modern and thorough treatment of the topic in the Bryozoa is yet to be made and would be well worth doing.

RESEARCH ON FISH

When Smitt became curator of the Division of Vertebrates at the Riksmuseum in 1871, he transferred his interests in deriving principles for understanding evolutionary pathways to a group about as far from colonial invertebrates as possible—fish. His emphasis was again on understanding growth. In particular, he used statistical analysis. The method is well illustrated as follows:

First, for the study of individual differences I have chosen the family of the salmon in which the differentiation of species has always been difficult, particularly because of their great variability, which will never be scientifically understood until one is able to discover the rules which govern it. And these rules, in a memoir titled "Riksmuseets Salmonider," published 10 years ago, I have arranged under three principal points: know the changes accompanying either the growth of individuals, or the differentiation of sexes, or the variations in the environment (Smitt, 1896a: p. 236).

As an example, he showed how apparently continuous variation in a widespread European species could in fact be separated into four clearly distinct geographic races by comparing the relative growth of parts of the body, within a single sex.

By emphasizing the partitioning of apparently continuously variable characters into their correct growth sequence, and taking sexual differences into consideration, Smitt was led to consider the way in which these species had arisen. In considering again the salmon, he wrote:

Thus I have given, I believe, the demonstration of the manner of differentiation of these species, which at one time had a common origin. A species is arrested in its development at a more or less lower stage, or else according to the influence which sexual or perhaps geographic differences have exercised, and it is turned aside from the path of development of other species—of which there are several—in pursuing its own course (Smitt, 1896a: p. 237).

For distinguished exposition and significant discovery, the salmon monograph won the Letterstedt Award of the Swedish Academy in 1879.

Smitt's best known work on fishes is his extensive revision of *Scandinavian Fishes* (1892–1895), issued in both Swedish and English editions. In it he included a large amount of biological information on the life of the fish, and he applied the principles discussed above, much as he had much earlier applied the principles of growth in bryozoans in differentiating taxa in the Scandinavian bryozoan fauna. *Scandinavian Fishes* remains the most important summary of marine fishes for that part of the world.

The application of taxonomic data to practical problems is another theme in Smitt's work. He had a strong practical interest in fisheries and encouraged the introduction of drift-nets into the Bohuslän herring fishery. As is almost inevitable for a person in a position of political power, arguments arose as to the proper course of action. Since Smitt encouraged the application of fisheries data, he was attacked in due course. The argument is described as "bitter, with many hard words on both sides" in the official Swedish Academy of Science biography of Smitt. Finally, Smitt represented the Swedish government at several international fisheries meetings.

OTHER RESEARCH AND RECOGNITION

In the official biography of Smitt published by the Swedish Academy of Sciences, it is reported that Smitt was extremely interested in whales for many years. He brought together a large number of skeletons at the Riksmuseum, etc. and was preparing a large volume on them. In this regard he is like another great student of the Bryozoa, Sir Sidney F. Harmer (1862–1950), who was at the British Museum. Like Smitt, he published extensively on Bryozoa but relatively little on whales.

F. A. Smitt also gained considerable recognition as a writer for the general scientific reader. He is best known in this regard for the 1876 book on the developmental history of higher animals, and the 1882 book on vertebrates. Also acclaimed was the translation from German into Swedish of A. E. Brehm's books on animal biology. This publication, which required three volumes, won for Smitt the Swedish Academy Prize for translation in 1885. Smitt was also active as secretary for the association for the promotion of silkworm culture in Sweden.

Smitt was elected to membership in the Swedish Academy of Sciences, and in the Central Chamber of Agriculture in 1875, and in the Göteborg Society of Science and Literature in 1878. He was also elected a Knight of the Order of the North Star in 1880, and an officer of the Royal Portuguese Order of St. James of the Sword.

Since 1947 F. A. Smitt's manuscripts have been kept at the Swedish Academy.

BIBLIOGRAPHY

1. 1861. "Sur les éphippies des Daphnies." *Nova acta Regiae societatis scientiarum upsaliensis. Kongl. Vetenskapssocieteten i Upsala,* ser. 3, 3: pp. 37–50, pls. 4–5.

2. 1863. Bidrag till Kännedomen om Hafs-Bryozoernas utveckling. Upsala Universitets Årsskrift. 40 p. [Contribution to the knowledge of development in marine Bryozoa.] An abstract of this article written by Alexander Agassiz appeared in two journals. The wording is identical in both. The references are: "Polymorphism among Bryozoa." *American Journal of Science and Arts,* ser. 2, 47. (1866a): p. 134; "Polymorphism among Bryozoa." *American Journal of Conchology* 3 (1866b): p. 105.

3. 1864. *Några drag ur Bryozoernas (mossdjurens) lif. Afhandling som jemte theser, kommer att offentligen försvaras inför högvördiga domkapitlet i Lund* (VIII-31 p. Stockholm. P. A. Norstedt & S:r.). [A few aspects of Bryozoan (moss-animals) life. A dissertation which, together with theses, will be publicly defended before the right reverend Chapter of the Diocese of Lund.]

4. 1865a. "Kritisk förteckning öfver Skandinaviens Hafs-Bryozoer. I." *Öfversigt af Kongl. Vetenskaps-Akademiens Förhandlingar* 22, 2: pp. 115–142; pl. 16. [Critical catalogue of the Scandinavian Marine Bryozoa. I.]

5. 1865b. "Om Hafs-Bryozoernas utveckling och fettkroppar." *Öfversigt af Kongl. Vetenskaps-Akademiens Förhandlingar* 22, 1: pp. 5–50; pls. 1–7. [On the development and fat-corpuscles of Marine Bryozoa.]

6. 1867. "Kritisk förteckning öfver Skandinaviens Hafs-Bryozoer. II." *Öfversigt af Kongl. Vetenskaps-Akademiens Förhandlingar* 23(Supplement): pp. 395–533; pls. 3–13. [Critical catalogue of the Scandinavian Marine Bryozoa. II.]

7. 1868a. "Kritisk förteckning öfver Skandinaviens Hafs-Bryozoer. III." *Öfversigt af Kongl. Vetenskaps-Akademiens Förhandlingar* 24, 5: pp. 279–429, pls. 16–20. [Critical catalogue of the Scandinavian Marine Bryozoa. III.]

8. 1868b. "Bryozoa marina in regionibus arcticis et borealibus viventia." *Öfversigt af Kongl. Vetenskaps-Akademiens Förhandlingar* 24, 6: pp. 443–487. [Marine Bryozoa Living in Arctic and Northern regions.]

9. 1868c. "Kritisk förteckning öfver Skandinaviens Hafs-Bryozoer. IV." *Öfversigt af Kongl. Vetenskaps-*

Akademiens Förhandlingar 24(Bihang): pp. 1–230, pls. 24–28.
[Critical catalogue of the Scandinavian Marine Bryozoa. IV.]

10. 1870. "De senaste årens undersökningar om hafsfaunans gräns mot djupet." *Framtidens tidskrift för fosterländsk odling.* Stockholm 3: pp. 335–349.
[Concerning the last year's investigations of the limit of the marine fauna toward the depth [of the ocean.]]

11. 1871. "Kritisk forteckning öfver Skandinaviens Hafsbryozoer. V." *Öfversigt af Kongl. Vetenskaps-Akademiens Förhandlingar* 28: pp. 1115–1134; pls. 20–21.
[Critical catalogue of the Scandinavian Marine Bryozoa. V.]

12. (with C. T. VON SIEBOLD) 1872a. "Bemerkung zu Dr. H. Nitsche's Beiträgen zur Kenntniss der Bryozoen." *Zeitschrift für wissenschaftliche Zoologie.* 22: pp. 281–282.
[Observations on Dr. H. Nitsche's contributions to the knowledge of Bryozoa.]

13. 1872b. "Remarks on Dr. Nitsche's researches on Bryozoa." *Quarterly Journal of Microscopical Science,* n.s., 12: pp. 246–248.

14. 1872c. "Floridan Bryozoa, collected by Count L. F. de Pourtales. Part I." *Kongl. Svenska Vetenskaps-Akademiens Handlingar* 10, 11: 1–20, pls. 1–5.

15. 1873. "Floridan Bryozoa, collected by Count L. F. de Pourtales. Part II." *Kongl. Svenska Vetenskaps-Akademiens Handlingar* 11, 4: pp. 1–83, pls. 1–13.

16. 1875. "Storkarne." *Läsning för folket.* 7: pp. 175–186.
[The storks. Readings for the people.]

17. 1876. *Ur de högre djurens utvecklingshistoria. Åtta populära föreläshingar* (Stockholm, P. A. Norstedt & S: r.) 276 p.
[Concerning the evolutionary history of higher animals.]

18. 1878a. "Recensio systematica animalium Bryozoorum, quae in itineribus, annis 1875 et 1876, ad insulas Novaja Semlja et ad ostium fluminis Jenisei, duce Professore A. E. Nordenskiöld, invenerunt Doctores A. Stuxberg et H. Théel." *Öfversigt af Kongl. Vetenskaps-Akademiens Förhandlingar* 35, 3: pp. 11–26.
[A systematic survey of the Bryozoan animals which Dr. A. Stuxberg and Dr. H. Théel found in trips made in the years 1875 and 1876 under the leadership of Professor A. E. Nordenskiöld, to the Island of Novaya Zemlya and to the month of the Jenisei River.]

19. 1878b. "Recensio animalium Bryozoorum e mari arctico, quae ad paeninsulam Kola, in itinere anno 1877, duce H. Sandeberg, invenit F. Trybom." *Öfversigt af Kongl. Vetenskaps-Akademiens Förhandlingar* 35, 7: pp. 19–32.
[A survey of the Bryozoan animals from the Arctic Sea which F. Trybom found on the Kola Peninsula in a trip made in the year 1877 under the leadership of H. Sandeberg.]

20. 1878c. "Ueber Balaenopter sibbaldi, Gr." *Zoologischer Anzeiger.* 1: pp. 365–366.

21. 1881a. *Föredrag i zoologi vid Vetenskapsakademiens högtidsdag den 31 mars 1881* (Stockholm, R. Wall), 16 p.
[Lecture in zoology at the annual festival day of the Academy of Science, March 31, 1881.]

22. 1881b. "Om fiskrikedomen (och om de åtgärder, som böra och lampligen kunna vidtagas för befordrandet af svenska storsjöfisket). Anföranden i Landtbruks-Akademien 11 April 1881 af C. Rydquist, Fr. Smitt, A. J. Lyth, A. Cederström Jr., och J. Arrhenius." *K. Landtbruks Akademien, Handlingar och Tidskrift* 20: pp. 135–164.
[On the abundance of fish (and on the protective measures that ought and could be taken to promote the oceanic fishery). Speeches in Landtbruks-Akademien,

April 11, 1881, by C. Rydquist, Fr. Smitt, A. J. Lyth, A. Cederström Jr. and J. Arrhenius.]

23. 1882a. *Inom eller utom skärs. No. 1* (Stockholm, P. A. Norstedt & S: r.), 8 p.
[Inside or outside rocky islets.]

24. 1882b. *Inom eller utom skärs. No. 2* (Stockholm, P. A. Norstedt & S: r.), 4 p.
[Inside or outside rocky islets.]

25. 1882c. "Description d'un Hareng hermaphrodite." *Archives de Biologie* 3: pp. 259–275, pl. 11.
[Description of an hermaphrodite Herring.]

26. 1882d. "Ryggradsdjurens geologiska utveckling och slägtskapsförhållanden." *Ur vår tids forskning* 29: 83 p.
[The geological development and mutual affinities of the vertebrates. Research of our times.]

27. 1882e. "Schematisk framställning af de i Riksmuseum befintliga laxartade fiskarnes slägtskapsförhållanden." *Öfversigt af Kongl. Vetenskaps-Akademiens Förhandlingar* 39, 8: pp. 31–40.
[Schematic description of salmon-type fishes preserved in the Riksmuseum and their mutual relationships.]

28. (Translator). 1882f. Däggdjurens lif [by A. E. Brehm]. Upplaga 2. Autoriserad öfversättning och bearbetning af F. A. Smitt och J. Lindahl, *granskad* och efter orig: s 2: dra upplaga redigerad af F. A. Smitt (Stockholm, Emanuel Girons Förlag. Ivar Haeggströms Boktryckeri), xx + 617 p.
[Life of Mammals. 2nd Edition. Authorized translation and revision by F. A. Smitt and J. Lindahl, corrected according to the 2nd edition of the original, edited by F. A. Smitt.]

29. 1883a. *The Swedish Fisheries. International Fisheries Exhibition, London. Special Catalogue* (W. Clowes and Sons, Ltd., London, England), 20 p.
[Evidently also published in Stockholm, by Kungliga Boktryckeriet, pp. 178–184.]

30. 1883b. Sveriges deltagande i Internationela Fiskeriutställningen i London 1883 (Stockholm, P. A. Norstedt & S: r.), 14 p.
[Sweden's participation in the International Fisheries' Exhibition in London, 1883.]

31. 1883c. *Underdånigt betänkande med förslag till ny fiskeristadga med mera, afgifvet der 3 mars 1883 af särskildt i döder förordnade kommitterade,* (Per Ehrenheim, J. Arrhenius, F. A. Smitt. . .) (Stockholm, Kunglig Boktryckeri), 159 p.
[Humble report with a proposal for a new regulation of the fisheries (etc.), given the 3rd of March, 1883, by individuals of an appointed committee (Per Ehrenheim, J. Arrhenius, F. A. Smitt. . . .).]

32. (Translator). 1884a. Fåglarnas lif [by A. E. Brehm]. Upplaga 2 Autoriserad öfversättning och bearbetning af F. A. Smitt (Stockholm, Emanuel Girons Förlag. Ivar Haeggströms Boktryckeri), xix + 747 p.
[Life of Birds. 2nd Edition. Authorized translation and revision by F. A. Smitt.]

33. 1884b. "Om åtgärder för befrämjande af länets hafsfisken. Föredrag vid Hushållning—sällskapets sammankomst den 31 januari 1884." *Tidning för Stockholm läns hushållning—sällskapet,* p. 49–56.
[On protective measures to advance marine fishing in the district. Lecture at the meeting of the Management-Society, January 31, 1884.]

34. 1886. "Kritisk förteckning öfver de i Riksmuseum befintliga Salmonider." *Kongl. Svenska Vetenskaps-Akademiens Handlingar* 21, 8: pp. 1–290, 6 pls., and 13 tables.
[Critical catalogue of the salmon preserved in the Riksmuseum.]

35. (Translator). 1887. *De Kallblodiga ryggradsdjurens lif* [by A. E. Brehm]. Upplaga 2. *Autoriserad öfversätt-*

ing och bearbetning af F. A. Smitt (Stockholm, P. A. Norstedt & S:r.), vii + 468 p.
[Life of Cold-blood Vertebrates. 2nd Edition. Authorized translation and revision by F. A. Smitt.]

36. 1888a. "Om Sillrasernas betydelse." *Kongl. Svenska Vetenskaps-Akademiens Handlingar, Bihang* 14, part 4, 12: pp. 1–18.
[Concerning the importance of races of herring.]

37. 1888b. "Om Trachypteridernas stjertfena." *Biologiska föreningens Förhandlingar. Stockholm* 1: pp. 17–21.
[On the tail fin of trachypterids.]

38. 1889. *Några ord om det begärda anslaget till Riksmusei etnografiska samling* (Stockholm, Kungliga Boktryckeriet), 10 p.
[Some remarks about the requested grant for the ethnographic collection of the Riksmuseum.]

39. 1892–1895. *Skandinaviens fisker.* Målade af W. von Wright. Beskrifna af B. Fries, C. U. Ekström och C. Sundevall. 2: dra upplagan. Bearbetning och fortsättning af F. A. Smitt. Stockholm, P. A. Norstedt & S:r. 4:0. part 1. Text: p. 1–566 + viii; pl. 1–22, 22A, 23–27. part 2. Text: p. 567–1239; pl. 27A–53 + iii.
[See entry no. 40 in this bibliography.]

40. 1892–1895. *A History of Scandinavian Fishes,* by B. Fries, C. U. Ekström, and C. Sundevall, with colored plates by W. von Wright. Translated into English by D. L. Morgan, from the 2nd revised and completed edition by F. A. Smitt. (2 pts., London, P. A. Norstedt & S:r), Part 1. Text: p. 1–566 + viii; plates 1–22, 22A, 23–27. Part 2. Text: p. 567–1240, plates 27A–53 + iv.

41. 1895. "Bedrifves det bohuslänska sillfisket på det för landet fördelaktigaste sätt?" *Skogs-och lantbruks-akademiens Handlingar och Tidskrift. Stockholm* 34: pp. 55–64.
[Is the Bohusland herring fishery being managed in the most profitable way for the country?]

42. 1896a. *La filiation des espèces d'animaux. Compte-Rendu des Séances du Troisième Congrès International de Zoologie, Leyde, 16–21 Septembre, 1895* (Leyde, E. J. Brill, Publisher), p. 235–238.
[The relationship between animal species.]

43. 1896b. "On the Habitat of *Gobius elapoides* G[ün]th[e]r." *Annals and Magazine of Natural History,* ser. 6, 18: p. 196.

44. 1897a. "Poissons de l'expédition scientifique à la Terre de Feu. I." *Kongl. Svenska Ventenskaps-Akademiens Handlingar, Bihang,* n.s., 23, part 4, 3: pp. 1–37, 3 pls.
[Fish of the scientific expedition to Tierra del Fuego. I.]

45. 1897b. *P. M. [Beträffande Riksmusei etnografiska samlingar]* Stockholm, (Iduns tryckeri-aktiebolag), 5 p.
[P. M. (Concerning the ethnographic collections of the Riksmuseum.)]

46. 1898a. "Poissons de l'expédition scientifique à la Terre de Feu. II." *Kongl. Svenska Venskaps-Akademiens Handlingar, Bihang,* n.s., 24, part 4, 5: pp. 1–80, 6 pls.
[Fish of the scientific expedition to Tierra del Fuego. II.]

47. 1898b. "Nyare tiders Darwinism. Föredrag vid Vetenskapsakademiens årshögtid den 31 mars 1898. Stockholm." *Aftonbladets tryckeri.* Ur Aftonbladet, May 7, 1898; supplement.
[Darwinism in recent times. Lecture at Vetenskapsakademien's annual festival day, March 31, 1898. Stockholm. From the newspaper Aftonbladet, May 7, 1898, supplement.]

48. 1899a. "*Phoca caspica* and *Phoca groenlandica.*" *Annals and Magazine of Natural History,* ser. 7, 4: pp. 339–341.

49. 1899b. "Preliminary Notes on the Arrangement of the Genus *Gobius,* with an Enumeration of its European Species." *Översigt af Kongl. Vetenskaps-Akademiens Förhandlingar* 56, 6: pp. 543–555.

50. 1900. "On the Genus *Lycodes.*" *Annals and Magazine of Natural History,* ser. 7, 5: pp. 56–58.

51. 1901. "Poissons d'eau douce de la Patagonie recueillis par E. Nordenskiöld, 1898–1899." *Kongl. Svenska Vetenskaps-Akademiens Handlingar, Bihang* 26, part 4, 13: pp. 1–31, 4 pls.
[Fresh water fish of Patagonia collected by E. Nordenskiöld in 1889–1899.]

52. 1901. "On the Genus *Lycodes.*" *Kongl. Svenska Vetenskaps-Akademiens Handlingar, Bihang* 27, part 4, 4: pp. 1–45, 1 pl.

53. A great many articles under the signature F. A. S. were written for the *Nordisk Familjebok* [Nordic Family Encyclopedia].

F. A. SMITT: HIS LATINITY AND THIS TRANSLATION

In his "Bryozoa Marina" Smitt uses a good modern scientific Latin, that is, a language with the inflections and syntax of classical Latin and with many new words formed on the pattern of classical Latin. (We mean by classical Latin the Latin of the period down to *ca.* A.D. 500, by medieval Latin that from *ca.* A.D. 500 to *ca.* A.D. 1500, by modern Latin that from *ca.* A.D. 1500 to the present.) Smitt's Latin seems fluent and is usually correct and free of ambiguities.

The following difficulties or oddities should be pointed out.

VOCABULARY

There seem to be two nouns each of which Smitt uses in two senses; the correct meaning can be determined only by the context. (1) *Apertura* is apparently used either for "orifice" or for "orifice plus uncalcified frontal"; in the latter case we translate by the current technical term "opesium." (2) *Area* seems to mean either "region" (in combination with *aperturae*) or "frontal area" (of course, we sometimes find the full phrase *area frontalis; cf.* p. 470L, p. 22E: ". . . nulla majore area frontali continua zoooecii relicta. . ."; ". . . with no additional continuous frontal area of the zooecium remaining . . ."). We have always translated *area aperturae* by "region of the opesium."

INFLECTION

In their English context we, of course, cite (Latin) scientific names in the nominative case, with the names of families ending in *-idae* and those of suborders in *-ina.* We have occasionally added in brackets [ ] some exegesis or the modern counterpart of a scientific name used by Smitt but no longer current.

Of Smitt's forms the following are noteworthy:

(1) Pp. 461L, 462L, 463L, 466L, 17E, 17E, 17E, 18E. He uses the genitive singular *Alectous* of *Alecto,* though the (Greek) ending *-ous* normally appears in Latin as *-us* (*-ūs*) (Neue, 1902: pp. 456–457).

(2) *Alterae* as a genitive or dative singular feminine, instead of *alterius, alteri,* is found occasionally in classical Latin (Neue, 1892: p. 539); Smitt regularly uses it. He also has twice on p. 482L (26E) the genitive

singular feminine *aliae*, which sometimes occurs in classical Latin, instead of *alius* (or the *alterius* that was substituted for it).

(3) P. 476L, 24E. *Dua* for *duo* (neut.), "two," may be a slip on Smitt's part or a misprint. It was considered a barbarism by Quintilian (1. 5. 15) unless it was in composition (*duapondo*, "two pounds"). But there are a few occurrences of it (e.g., *loca dua*) in inscriptions (usually of the late classical period) and the manuscripts of authors; *cf.* Neue (1892: p. 177) and Thesaurus, v. 5 (1910–1953: cols. 2241–2242). In any event, it clearly functions as the neuter nominative in Smitt, and the sense is clear: ". . . quare dua haec genera inter *Cyclostomata* olim locum habebant"; ". . . as a consequence these two genera formerly were placed among the Cyclostomata."

(4) P. 472L, 23E. *Extus.* We have taken this as an adverb: ". . . docet jam forma zoooecii extus rectangularis *Cellariae borealis* . . ."; "The external [lit., externally] rectangular form of the zooecium of *Cellaria borealis* shows. . . ." The form *extus* as either an adverb or an adjective seems not to have existed in Classical Latin (*cf.* Thesaurus, v. 5, 1910: col. 2093) but is readily understandable as an adverb made on the analogy of the adverbs *intus* and *subtus*; maybe Smitt himself is responsible for the creation. It seems to be a doubtful form in medieval Latin; we have found just one reference for it as an adjective meaning "foreign" or "alien," but as a variant reading in a text, in the new edition of Du Cange, v. 3 (1884: p. 383).

(5) Pp. 476–477L, 24E. *Nequat.* We have taken this as a misprint for *nequeat* (present subjunctive form of *nequeo*, "am unable"): "Clauditur vero citius hoc zoooecium et calcificatione progrediente ita obtegitur, ut forma primaria recognosci nequat"; "Now this zooecium is closed more quickly and is covered with an advancing calcification so that its primary form is not able to be distinguished."

P. 482L, 26E. *Poterint.* We have taken this as a slip of Smitt's or a misprint for *poterunt:* "Hos secundum modos duos evolutionis *Escharina* fortasse aliquando in duas series distribui poterint . . ."; "According to these two ways of evolution, *Escharina* perhaps will at some time be able to be distributed into two series. . . ."

(6) P. 472L, 23E. *Nostrius* seems to be a mistake of Smitt's (a form made on the analogy of *alterius, totius*, etc.): ". . . cujus formae nostrius faunae parum supra *Ctenostomata* elevatum stadium differentiationis tenent"; "Their forms [the forms of the family Flustridae] in our fauna have a stage of differentiation slightly above the Ctenostomata."

(7) P. 475L, 24E. Smitt's *poribus* is a mistake for *poris* (ablative plural). The meaning is clear enough: "Sed e poribus unus in fronte zoooecii fere medius. . ."; "But one of its pores, almost in the middle in the front of the zooecium. . . ." We find the correct dative plural

poris on the same page (475L) and the correct ablative plural, also *poris*, elsewhere.

(8) P. 475L, 24E. We have taken the form *propiarum* as a mistake for *propiorum* (masc., fem., and neut. gen. pl. of *propior, propius*, "nearer") and translated accordingly: "Praeterea apertura zoooecii transverse elliptica, magis minusve elevata, interdum aliquanto quadrangulata, affinitatem etiam *Discoporidarum, Celleporinis* longe propiarum, indicat haec familia. . ."; "In addition, the orifice of the zooecium of this family is transversely elliptical, more or less elevated, and sometimes somewhat quadrangular; this [characteristic] indicates also the affinity of the family Discoporidae which is much closer to [the suborder] Celleporina."

(9) *Quum,* regularly used by Smitt for the conjunction *cum* (on the problem of its meaning see below), is found in medieval manuscripts and modern printed texts. But it lacks ancient authority; *cf.* Neue (1892: p. 633, where Pohl's investigations are cited for its absence from inscriptions), and Buck (1948: p. 83).

WORD FORMATION

Smitt follows standard procedures in the formation of new words. If he uses new forms of compounds, they are on the analogy of either classical Latin items in their own right, so to speak, or as influenced by Greek, which was much more given to compounding. (Of course, as in the tradition of Latin scientific nomenclature, there are forms in Smitt that are almost purely Greek, e.g., those of *perigastricus*, p. 465L note 1, 486L.) We have not tried to establish just which forms or compounds were coined by him; some of those that we shall note below may have been created by earlier scientists or scholars in other fields. We merely list a few forms with suggestions as to classical prototypes.

(1) *Verticillatus*, "verticillate" (*verticillatis*, p. 471L, 22E), corresponds to ancient forms like *barbatus* and *hastatus; cf.* Buck (1948: p. 335).

(2) We have not found the abverb *labiatim*, "labially" (p. 472L, 23E), or *labeatim*, in dictionaries of classical or medieval Latin. But adverbs in *-atim* are common in Latin; *cf.* the long list in Neue (1892: pp. 549–564).

(3) Such forms as *proximalis* and *distalis* occur frequently. Many of them may not be found in classical Latin, but they can be justified on the analogy of something like *terminalis*, which is ancient.

(4) Compounds in *-formis* are characteristic of botanical and zoological Latin and abound in Smitt. He has *crustiformis, infundibiliformis, rostriformis, S-formis, Tata-formis*, etc. Such compounds are found in classical Latin (*cf.* the list in Gradenwitz, 1904: pp. 460–461) though they often seem somewhat artificial and were made up on the pattern of Greek compound adjectives ending in *-morphos*.

(5) The adjective *obconicus* (*obconicam*, p. 470L, 22E) is a modern formation, perhaps on the analogy of

oblongus. We translate by "rather conical" (*cf.* the meaning of *oblongus* in classical Latin: "rather long") and not "inversely conical" (*cf.* "obconical" in the *Oxford English Dictionary;* note there also *ob-* pref. 2, where the *ob-* of Latin *obverse,* rather than the analogy of *oblongus,* is suggested for the etymology of scientific items like "obtriangular," "triangular with the apex downward").

SYNTAX

The following details should be pointed out:

(1) Many of Smitt's dependent clauses introduced by *quum* (= *cum*) are straightforward enough; for instance, the verb is in the subjunctive and *quum* means "since," or the verb is in the indicative and *quum* means "when." But there are other places where the verb is in the indicative and *quum* seems to mean something like "since." Here the clauses are of the explicative type (historically, a development of the temporal); *cf.* Hofmann (1965: pp. 619, 624). Such meanings as "in that" and "on the ground that" for the *quum* (*cum*) of these clauses are found in Latin grammars written in English. We have translated by "since" twice and by "in that" twice; we have made a relative clause beginning with "in which" once; three times we have turned the dependent clauses into sentences, without the *quum,* and allowed the causal relationship to be inferred. The eight instances of *quum*-clauses which we have construed this way are the following: p. 464L, 18E: "Sed etiam in his formis transitum e stadiis simplicioribus videmus, quum progressum calcificationis in omnibus fere stirpibus ex apice ad basin persequi possumus ...."; "But even in these forms we see a transition from the simpler stages. We are able to follow the progression of calcification...." P. 465L, 18E: "... zoooecia ... eademque fere compositione quam praebent illa animalia, quum gemma Bryozoarii communis vesicae formam tenet ..."; "The zooecia have almost the same structure as those animals in that the common bud of the bryozoarium has the form of a bladder...." P. 470L, 22E: "... quare infimum evolutionis gradum vulgo tenent, quum e stadiis cum *Ctenostomatibus* congruentibus eo ipso modo evadunt *Chilostomata,* ut partem proximalem pone aream aperturae construant ..."; "Therefore they commonly maintain the lowest grade of evolution in that the Cheilostomata evolve from stages corresponding to the Ctenostomata in just such a way that...." P. 472L, 23E: "... optima nobis exempla, quam parvi haec sit momenti, species *Flustrae* praebent, quum et crustiformes et in ... formam erectae crescunt, simplices vel ex duobus zoooeciorum stratis compositae"; "... the species of *Flustra* offer the best examples to us to show how insignificant this is. Both the simple ones and those formed of two layers of zooecia grow encrusting and erect into the form of...." P. 478L, 25E: "... quarum differentiarum vim jam de *Membraniporis* adnotavimus, majorem quum evolutionis gradum indicat

forma acuminata mandibulae (i.e. operculi) avicularii"; "We have previously noted the significance of these differences in *Membranipora* in which the pointed form of the mandible (i.e., the operculum) of the avicularium indicates a higher grade of evolution." P. 478L, 25E: "Quorum quum forma et in his speciebus plana est, facilius e praecedentibus distinguuntur avicularii mediani mandibula ..."; "Since the form of these [zooecia] is flattened also in these species, they [the species] are distinguished more easily from those preceding by the mandible of the medial avicularium ...." P. 479L, 25E: "Quarum tamen vim opinatam haud de his valere credere possumus, quum indicat operculum ... avicularii actionem"; "Nevertheless, we cannot believe that that fancied interpretation regarding the *cellulae fertiles* is correct since the operculum indicates the action of an avicularium...." P. 481L, 26E: "Duobus enim modis constrictiones suas efficit, quum aut in annulo transverso clauduntur zoooecia ..."; It forms constrictions two ways. The zooecia are either enclosed in a transverse ring...."

Twice Smitt seems to use *quum* with a primary-tense subjunctive verb in an iterative sense; see for this Hofmann (1965: p. 624). We have translated the *quum* each time by "when": P. 469L, 21E: "Incidit vero magna illa difficultas, quum species quaedam in stadio uno vel minus evoluto descripta sit, veram ejus affinitatem censendi"; "But when a given species has been described in one stage, or in a less-evolved stage, a well-known, major difficulty arises, namely of assessing its true relationship." P. 474L, 23E: "Fit tamen hic quod saepe, quum series plures formarum e primordio simili progrediantur, ut multis modis invicem se tangant"; "But there occurs here what is often found when several series of forms proceed from a similar origin, namely that they may look like each other in many ways."

(2) One conditional sentence (actually a conditional sentence inside an *ut*-clause of result) has been bothersome: P. 485L, 27E: "... ut, augente magnitudine aviculariorum lateralium ... plures species distinguendas censeremus, si formas intermedias et differentias evolutionis ignoraverimus." We have translated: "... so that, with the increasing size of the lateral avicularia ... we would think that more species ought to be distinguished, if we did not know about the intermediate forms and the differences in their evolution." Thus we have taken *ignoraverimus* for the *ignoraremus* which would be expected. Smitt's perfect subjunctive here (*ignoraverimus*) may be a confused recollection of the periphrastic perfect subjunctive for the pluperfect subjunctive in the conclusion of a past contrary-to-fact conditional sentence depending on a sentence requiring the subjunctive; see Gildersleve (1895: p. 387, especially the example from Livy, 26.10.7).

(3) In five instances we have taken a perfect-tense form to be gnomic (the perfect of a general truth) and so translated by a present. An example of this from

classical Latin would be the perfect *receperunt* (matching the present *fovent*) in Seneca (1969: 18.15): "Sic ignis non refert quam magnus sed quo incidat; nam etiam maximum solida non receperunt, rursus arida et corripi facilia scintillam quoque fovent usque in incendium"; "So it does not matter how big a fire is, but where it happens, for solid objects do not yield to even the biggest fire, whereas dry and inflammable ones nurse even a spark into a conflagration" (translation adapted from Seneca, 1969: p. 69). In four of the examples in Smitt the verb is a form of *praebere*, "offer, provide, share" (to which he is greatly addicted); in the remaining one, a form of *habere*, "have." The underlying idea would be: a given species has such and such characteristics because all the examples that we know of it have presented us with those characteristics. The five passages in Smitt are as follows. P. 471L, 22E: "Species generis *Cellulariae* evolutionem coloniae praebuerunt e primo zoooecio magis minusve *Tati*-formi . . . procedentem"; "Species of the genus *Cellularia* illustrate the evolution of the colony proceeding from a first zooecium which is more or less a tata-form." P. 486L, 28E: "Et *Retepora notopachys*, quam typicam censemus, in mari mediterraneo oooecia semper rimata praebuit"; "And *Retepora notopachys*, which we consider to be typical, in the Mediterranean Sea has ooecia which always have clefts." P. 487L, 28E: "*Retepora beaniana*, nostras, avicularium minusculum ad aperturam zoooeciorum laterale quamvis rarum, mandibula triangulari clausum, praebuit, quare transitus ad *R. cellulosam* jam clarius aperitur."; "*Retepora beaniana*, in our regions, has a minuscule lateral avicularium at the zooecial orifice, although it is rare, closed by a triangular mandible, so that the transition to *R. cellulosa* is now more evident." P. 487L, 28E: "*Celleporaria incrassata* evolutionem coloniae inde a primo zoooecio Tatiformi praebuit"; "*Celleporaria incrassata* shows the evolution of a colony from the very first tata-form zooecium." P. 487L, 28E; "Alia vero colonia hoc zoooecium jam erectum, celleporinum habuit"; "But another colony has its zooecium built up in a manner like a celleporine."

(4) P. 465L, 18E: ". . . palaeontologis est de hac quaestione responsum dare"; ". . . it is for paleontologists to provide an answer to this question." The meaning is clear enough, though a more idiomatic classical Latin would be . . . *palaeontologorum est* . . . or even a quite different construction such as *palaeontologi de hac quaestione responsum dent.*

(5) P. 467L, 19E: ". . . adnotationes, quae supra de variationibus fecimus . . ."; ". . . the observations we have made above concerning the variations. . . ." We have taken *quae* to be a lapse on the part of Smitt or the type-setter for *quas*.

(6) P. 472L, 22E: "Hunc evolutionis modum praesertim memoratu dignum est. . . ."; "It is especially necessary to consider this pattern of evolution. . . ." Again the meaning is clear, though the Latin construction is incorrect. Smitt seems to have conflated two constructions: *Hunc evolutionis modum praesertim memorare dignum est*, and *Hic evolutionis modus praesertim memoratu dignus est.*

(7) P. 487L, 28E: ". . . tam arcte . . . ut familia *Eschariporidarum* familiam quam fecimus *Porinidarum* amplectetur"; "so closely that the family Eschariporidae embraces the family which we have made of the Porinidae." *Amplectetur* (fut. indic.) instead of *amplectatur* (pres. subjunc.) in an *ut*-clause of result is either a slip on Smitt's part or a misprint.

STYLE

Smitt's Latin style is direct and straightforward and presents few flourishes. One embellishment is the whole penultimate paragraph, where he dilates on the laws of evolution. Note the alliterating doublet *interruptum et incertum*, "interrupted and uncertain" (P. 486L, 28E), *cf. exstinctas externasque*, "extinct and peculiar" (P. 480L, 26E). Note also what we have taken as an instance of hendiadys: *variationibus et calcificatione*, "by variations in calcification" (P. 486L, 28E, literally "by variations and calcification"; *cf.* for an example from classical Latin *pateris et auro*, "with cups of gold" or "with golden cups," literally "with cups and gold").

Long sentences seem characteristic of many Latin texts from classical times onwards, and Smitt has a fair number of them. In fact, he shows in his use of them his facility in Latin, and he does not lose his syntactic thread and fall into anacolutha. But we have broken up many of his long sentences. To do this, we thought, was more in keeping with modern English and would make it easier for the reader to follow the thread of his argument.

We have also added headings to the different sections of "Bryozoa Marina" to make it easier to find where different topics are discussed.

MARINE BRYOZOA LIVING IN ARCTIC AND NORTHERN REGIONS

Presented to the Academy on 12 June, 1867

INTRODUCTION

COLLECTIONS

The collections which we have been able to use were obtained from the following sources:

Professor S. Lovén, who, as early as 1832–1836, possessed in the Swedish Museum at Stockholm and in his notes many new species which were described and fully characterized. He added to their number during expeditions to the western shores of Sweden, and in 1836–1837 to Finmark and to the Spitsbergen Islands;

Professor V. Lilljeborg, who gave to the Uppsala Museum many species of this group that were collected during his expeditions to Grip and Kristinsund in Norway;

Professor O. Torell, who deposited very large collections, which also included these animals, in the Swedish Museum at Stockholm, which he collected during his journeys to Iceland in 1857, to the Spitsbergen Islands in 1858, and to Greenland in 1859;

The Swedish expeditions to the Spitsbergen Islands during the years 1861 (Goes, Malmgren, Smitt) and 1864 (Malmgren);

A. W. Malm, a keeper at the Göteborg Museum, whose careful collecting at Bohuslän [Sweden, 57° 51' N, 12° 02 E] included this branch of zoology;

Baron E. Uggla, who gave to the Swedish Museum at Stockholm some very beautiful and rare specimens of Bryozoa which were attached to *Oculina* and *Gorgonia* that had been raised out of the depths by fishermen;

Doctor A. Boeck, a Norwegian, who generously offered us his excellent note books and illustrations;

Moeller, inspector of Greenland, who had investigated that area thoroughly; his collections in the Danish Museum at Copenhagen were kindly presented to us by Professor Jap Steenstrup and Doctor C. Lükten;

Furthermore Professor Lacaze-Duthiers and Professor d'Archiac and Doctor Fischer kindly showed to us at Paris the type specimens of Lamarck and d'Orbigny for determining synonyms. We have listed [in the following table] only those places of origin which these museums provided, or which we ourselves have seen. Therefore we invite others to complete the columns, especially that of Greenland. [In the original article, the occurrence of each species is indicated for southwestern Scandinavia, Finmark, Spitsbergen Islands, and Greenland. This table is recast here as tables 1–3 at the end. It is provided as a summary of the classification used in the text.]

METHODOLOGY

We wish to add to this account, so that it is better understood, certain observations about our method. In several places one may see that we enumerate forms as varieties of species which are listed as species by other writers on this subject. And concerning the divisions even of the higher classifications we offer several new interpretations. The basic question of classification depends upon correctly evaluating the nature of *species*. Often species vary and are joined together by variations in such a way that the distinguishing features are scarcely to be found, and the limits of even the broader divisions may become uncertain. Concerning the Bryozoa, this uncertainty has created great confusion. In order that this situation may be resolved, we shall try to demonstrate that the differences of these species are due to their evolution and that they can be explained for the most part by that principle.

If we seek the correct meaning of *species*, then the higher animals and those which maintain a more or less fixed shape from their earliest stages of a free life make this investigation more difficult because they commonly show minor and individual variation, although the species are very often more easily distinguished. However, if we examine the animals which live united in colonies, whose genealogy of individuals is therefore not to be disputed, we shall be able to investigate more easily the laws and reasons for variation. In this respect the Bryozoa will be of the highest assistance because great variations occur in their colonies. We can follow the variations when they are arranged together in series, from the simplest stages up to the most highly evolved forms. This is just as valid for individuals of species and for individuals of colonies, as it is for distinct species. Therefore the differentiation of a colony shows a series of variations that progresses in the same way that a series of differences progresses among species.

CTENOSTOMATA

The Ctenostomata have the least highly evolved skeleton. They are always membranaceous or fleshy. Their colony, and its organs, are on a lower scale of development. Their zooecia are just as those of the higher Bryozoa with regard to budding.[1] [Superscript numbers refer to bold-face numbers in the appendix, p. 29.]

Although the Ctenostomata are connected by the evolution of forms, they are separated into two families [see table 1]. The first family, the Halcyonellidae, includes the lowest forms. These look like incrustations and are composed of hexagonal zooecia that are joined to each other. But from this stage, two directions evolve. One leads to forms similar to *Hippothoa* [of the Cheilostomata] through the breaking up of the colony; the other retains its zooecia joined together but builds a colony in the form of a fleshy stem. In both of these cases, the distal (English = oral) part of the zooecium is gradually elongated.[2]

In *Alcyonidium hirsutum* all of these variations are joined in such a way that the first zooecia of each colony retain the prototype form. The breaking up of the

TABLE 1

Class Bryozoa

Tribe 1. Infundibulata

Order 1. Ctenostomata
    Family 1. Halcyonellidae ·
      Genus *Alcyonidium*
        Subgenus *Halodactylus*
          Alcyonidium mytili
          Alcyonidium hirsutum forma linearis
          Alcyonidium hirsutum forma mammillata
          Alcyonidium hirsutum forma hirsuta
          Alcyonidium hirsutum forma
          Alcyonidium gelatinosum
        Subgenus *Cycloum*
          Alcyonidium parasiticum
          Alyconidium papillosum
          Alcyonidium hispidum

    Family 2. Vesicularidae

      Genus *Vesicularia*
        Subgenus *Valkeria*
          Vesicularia uva
          Vesicularia cuscuta
        Subgenus *Farrella*
          Vesicularia familiaris
        Subgenus *Avenella*
          Vesicularia fusca

TABLE 2

Order 1. Cyclostomata
Suborder 1. Radicellatina
    Family 1. Crisidae
      Genus *Crisia*
        Crisia cornuta
        Crisia producta
        Crisia eburnea
        Crisia denticulata

Suborder 2. Incrustatina
  Section 1. Tubulinea
    Family 1. Diastoporidae
      Genus *Diastopora*
        Diastopora repens
        Diastopora simplex
        Diastopora hyalinea forma obelia
        Diastopora hyalina forma latomarginata
        Diastopora patina forma typica
        Diastopora patina forma radiata
      Genus *Mesenteripora*
        Mesenteripora meandrina

    Family 2. Tubuliporidae
      Genus *Tubulipora*
        Subgenus *Idmonea*
          Idmonea atlantica
          Idmonea fenestrata
          Idmonea serpens
        Subgenus *Phalangella*
          Phalangella palmata
          Phalangella flabellaris
        Subgenus *Proboscina*
          Proboscina incrassata
          Proboscina fungia
          Proboscina penicillata

    Family 3. Horneridae
      Genus *Hornera*
        Hornera proboscina
        Hornera violacea
        Hornera lichenoides

    Family 4. Lichenoporidae
      Genus *Discoporella*
        Discoporella verrucaria
        Discoporella crassiuscula
        Discoporella hispida

  Section 2. Fasciculinea
    Family 1. Frondiporidae
      Genus *Frondipora*
        Frondipora verrucosa

    Family 2. Corymboporidae
      Genus *Corymbopora*
        Corymbopora fungiformis
      Genus *Coronopora*
        Coronopora truncata

    Family 3. Defrancidae
      Genus *Defrancia*
        Defrancia lucernaria

colony together with the elongation of a distal portion of the animal paves the way to [the development of] the family Vesicularidae. But the family Halcyonellidae presents yet another type of variation. Those three species which we have joined together under the subgeneric name *Cycloum* (Hass.) are composed of zooecia, whose outer surface is distinguished by papilli which are hollow, conical,[3] or even elongated into the form of a spine, and which are located at the boundaries[4] [of zooecia]. In *Alcyonidium hispidum* these spines vary in number so that only the distal ones are retained; these spines display the prototype of a [pattern of] variation that is very common in the evolution of the Cheilostomata.

The second family of Ctenostomata, the Vesicularidae, varies especially in the prolongation of the zooecia. Since the organs of digestion and reproduction are included in a raised portion of the zooecium, the proximal (English = basal) part is changed into a tubule[5] which together with corresponding parts of its neighbors forms a tubular root or stem. A transition of this type leads to the form of *Aetea* (of the Cheilostomata).

But at this point we can follow another line of reasoning. The cylindrical form of the zooecium also connects *Vesicularia* and *Aetea* with the Cyclostomata. Concerning *Aetea anguina*, Busk has already observed that the origin of its colony is derived from a hemispherical body of the same type from which the Cyclostomata proceed in their evolution.[6] These cyclostomes also have a zooecial skeleton perforated with pores as is seen in the Aeteidae.

CYCLOSTOMATA

In the Cyclostomata,* the first family [see table 2], the Crisidae,[7] furnishes us the best example of forms that progress from the simplest stages, where the zooecia are alternating in double series and are loosely joined together, all the way to the construction of a firmer stem whose zooecia are mostly joined together. The forms are connected in such a way that very often *Crisia eburnea*, in the lower portions of the stem, has the form of *Crisia cornuta*, whereas *Crisia denticulata* shows the same relationship to *eburnea*; moreover, we see so many forms that are intermediate between these that we are not able to distinguish them by characteristic traits. However, in many places one form or the other occurs so unvaryingly that it can easily be considered a well defined species. *Crisia eburnea* does not always pass through the stage of *Crisia cornuta*, nor does *Crisia denticulata* always at first resemble *Crisia eburnea*. We therefore conclude that the forms of this genus should be arranged according to the law of evolution in a sequence whose members, derived from a common origin, keep their own degree of evolution in such a way that each stage [in the sequence] seems to be a fixed species. The result of this is that we can propose as many species as we see more or less fixed stages.

This is also valid for the families which compose the [Cyclostome] suborder Incrustatina. But the more highly evolved they are, the more difficult it is to show a connection among the forms. This is because they have lived through long geological periods during which several intermediary forms have became extinct, and because the mode of origin is often obscured by the more intricate organization of the colony. Even though a large number of species have been described from this suborder, I wish to point out that almost no one has made observations about their evolution. The forms that we know, however, have to be arranged in several parallel series with the result that the corresponding stages often appear to be very similar. Because of this, considerable uncertainty arises concerning the synonyms. Nevertheless, we are certain that the very old genus *Alecto* indicates the first and simplest stages of evolution from which *Criserpia*[8] (of authors) and *Diastopora*[9] follow one path, whereas *Criserpia* (of authors) and *Tubulipora* follow another.[10] Up to now concerning the Horneridae we have seen no sure evidence of its origin, but our first form, *Hornera proboscina*,[11] shows an organization corresponding to *Pustulipora*, whose origin Hincks observed from the *Alecto* stage, which is very similar to the origin of *Proboscina*. We have been able to investigate the evo-

lution of the Lichenoporidae[12] in which *Discoporella* shows an early stage corresponding to *Tubulipora* (*Phalangella*) in which stage it too has primary lateral buds and scattered primary zooecia that very often form a more or less reniform colony.[13] The forms of Fasciculinea are more remote, but the section Tubulinea for the most part (e.g., *Phalangella flabellaris*, *Proboscina penicillata*) shows a transition to their structure. It seems reasonable—though I am still not able to show how— that that form of *Corymbopora*[14] which I have described under the name of *fungiformis* is an early stage of *Coronopora truncata*. From this line of reasoning we are able to guess about the manner of evolution of these forms too.

If we examine rather carefully the [other] families which we are enumerating, we shall see the same relationship among the forms which the family Crisidae shows. Thus *Diastopora repens* is broadened more and more out of the form of *Alecto*[15] and it arrives at the form of *Diastopora simplex*; and the latter is equivalent to an early stage of *D. hyalina*. *D. hyalina* and *D. patina* provide very memorable varieties that make a transition to the Tubuliporidae with the zooecia increasingly packed into radiating rows. It can no longer be doubted that *Mesenteripora* shows a sole characteristic method of growth that is of the same type as *Diastopora*. Indeed I have seen *Reticulipora nummulitorum* from the Mediterranean Sea[16] proceeding from a base similar to that of *Diastopora*. But since I have not as yet seen the base of *Mesenteripora*, I leave the question unanswered whether any of our arctic *Diastopora* is the origin of this genus. Often it happens that the early stages—under conditions whose causes we very often do not know—maintain as much control over their appearance as only the older stages generally have. Therefore it is not inevitable that all the younger stages develop into the older stages. Already we have noted this in *Crisia*, but it is even more evident in *Tubulipora*. First, we have been able to examine the continual evolution of *Idmonea atlantica* from the *Alecto* stage.[17] Its zooecia are brought together in transverse rows and increase both in size and in number[18] at the same time as the stem shows a more and more flattened dorsal surface.[19] It shows a transition of this type to *Idmonea serpens* so that Sars saw the reasons why it is referred to the same species as *I. serpens*. Nevertheless, in the stages of the same grade of evolution as *I. atlantica*,[20] *I. serpens* is always more flattened out and should be distinguished by its own name particularly because it illustrates another example of this type of evolution. Often it is able to retain longer the composition and mode of growth of *Criserpia*,[21] or having been composed in the same way is able to become erect from a base. Since this [ability to become erect from an encrusting base] occurs in *Phalangella* that commonly creep, and in *Proboscinia* that commonly are erect, it often happens that we find colonies or parts of colonies

---

* The name perhaps should be changed because of the order of fish, Cyclostomata.

from these three series which are hardly to be distinguished. We see that *Tubulipora lobulata* is none other than a kind of *Idmonea serpens* which extends itself in the manner of *Phalangella*, just as sometimes colonies of that species [*I. serpens*] emerging from the origin of the typical form [22] assume the appearance of *Phalangella*. We arrange the two series *Phalangella* and *Proboscina*, which develop parallel to each other, in the same way in which we arrange *Idmonea*, so that the one form approaches more and more in evolution the other of the same series. For this reason, the distinctions, just as among the Crisidae, can hardly be determined among the forms of these series. Thus we see that the family Tubuliporidae is composed of three series. But just as these proceed from a similar origin, so also the one can meet the other; therefore the generic differences are scarcely worth inquiring into.

The remaining families of the section Tubulinea, the Horneridae and the Lichenoporidae, have typical cancellous forms, which are strengthened and covered with additional calcification; this makes it harder for us to investigate the manner of origin. In addition, the typical *Hornera* has ooecia affixed on the back of the stem.[23] But even in these forms we see a transition from the simpler stages. We are able to follow the progression of calcification in almost all the stems from the apex to the base, although the ooecia differ in respect to the place of adhesion. Thus the apices of *Hornera lichenoides* [24] are composed of the same organization of zooecia as the whole stem of *Hornera proboscina*,[25] which carries the ooecia [26] on the front of the stem although it is joined so closely with *H. violacea*—whose ooecia according to Sars are attached to the back of the stem—that I do not dare separate them.

In the same manner the forms of the genus *Discoporella* are interrelated. Nevertheless their variations have so dissimilar an appearance that often they are believed to be quite distinct species. This is chiefly due to the stronger calcification and the buildup of the ooecia which hide the center of the colony (or centers, if the colony is enlarged many times by budding). A great variety of forms results when a new ooecium [27] is formed on top of another in series, after the new cancelli have been formed. But there are also zooecia of a dissimilar appearance, and thus we can distinguish three forms which, nevertheless, many intermediary forms join. Often *Discoporella verrucaria*, when it has grown to a greater length, copies the appearance of *D. crassiuscula* or of *D. hispida* and very often the early stages of *D. hispida* resemble *D. crassiuscula*.

We have already noted in the section Fasciculinea that their forms are more dissimilar from each other. Indeed it can come about that it seems that these families sometimes should be distributed in series proceeding from the Tubulinea, if we may conjecture that *Frondipora* corresponds to *Idmonea*, *Coronopora* to *Phalangella*, and *Defrancia* to *Proboscina*. However,

Filifascigera (d'Orb.) and *Reptofascigera* (d'Orb.) must also be noted which can indicate the characteristic evolution of the Fasciculinea right from the lowest stages. But since most forms of this section are extinct, it is for paleontologists to provide an answer to this question. If this answer is to be correct, then it ought to depend upon the proper understanding of budding, whose manner we have indicated in another discussion. For common buds generally arise at the margin [28] or apex [29] of the colony. These are divided by internal walls into zooecia, in the same way in which the Foraminifera grow. The zooecia have almost the same structure as those animals in that the common bud of the bryozoarium has the form of a bladder and is filled with material very similar to their protoplasm.** Thus the laws of growth of the Foraminifera and the Bryozoa seem similar—just as the forms of colonies also are, for example, in *Dactylopora*; it is worth recalling that Carpenter has observed the same relationship among their forms, although he pointed out that it is less explicit in regard to the lower and less well-known animals. The Bryozoa exhibit more distinct forms although they vary as much. One who wishes to investigate these should see the variations of the common bud, which is able to be differentiated in various ways as the growth of the colony proceeds, when the bud is extended from parts that are increasingly organized. Thus the relationship is brought about that the evolution of the colony leads towards a more multi-faceted organization.

In the case of our Cyclostomata, at least of those that are known, we have briefly clarified the relationships whereby this evolutionary reasoning arranges the forms. To give an example, the small differences by which the forms of *Alecto*, which are the origins of certain *Idmonea* and of *Proboscina*, are distinguished,[30] doubt-

---

** We have maintained the name *corpuscula adiposa* after Henle for this material. We will say emphatically, however, that it must be observed (Öfvers., 1865: p. 6) that the *corpuscula adiposa* is not the same thing as the *corpora adiposa* of insects, although this too can be converted through histolysis to a similar stage of composition, and although certain Vermes exhibit groups of free corpuscles which join the structure of both of these. But we did not wish to apply the name protoplasm for animals that are so highly organized since this material which fills the perigastric cavity of adult zooecia performs in these animals almost all of the functions of the lymphatic material, bile and blood of higher animals. Protoplasm remains from the egg after other organs have been formed and in the adult zooecium it grows by feeding. This nutrition in the bud is of the same type as that nourishment from which the new digestive organs arise after histolysis in adult zooecia, or as that which fills the ooecia, just as the organs of reproduction (or soon eggs and semen) also arise through a differentiation of this material. The Bryozoa show us enormous variations in this matter which seem to be of great importance if we wish to clarify methods of reproduction that are so diverse for these animals. For among these we often see several methods of reproduction at the same time in the same colony. For example, we have seen *Bugula flabellata* produce eggs both in the zooecium and in the ooecium affixed above this zooecium.

less indicate a common origin from which, however, even the forms presently living are able to diverge in very different ways sooner or later, just as we noticed in *Idmonea serpens*. Those forms that are lower in their evolutionary grade can exist in the manner of species, for instance as *Idmonea serpens* does when it piles up ooecia in the stage of *Criserpia*.[31] Or these lower stages are by no means necessary, just as we very often see *Idmonea atlantica* omitting the stage of *Alecto*. When we arrange the series of forms according to the mode of evolution, we are able to distinguish, after examining them, the different grades of evolution of the forms. I have already said that a certain form can exist at a stage that is advanced to a greater or lesser extent; thus we see that a certain variation of a certain grade exists in the manner of species, just as we noted in the case of *Idmonea fenestrata* or *Tubulipora lobulata*. Single series, although in a dissimilar grade, are able to vary in the same way, in such a manner that these variations are also more or less explicit. For example, *Diastopora patina* has a pan-shaped colony and has zooecia packed together into rows more frequently and at an earlier stage than does *Diastopora hyalina*. We have spoken about evolution, and about [species] characteristics whose significance we have indicated in accord with their evolution. But dissimilarities also occur which seem to indicate the power, so to speak, of species to diverge in their own way. As long as we are not able to explain the significance of these dissimilarities, there remain species that are quite distinct. For example, *Hornera frondiculata, Idmonea radians*, and *Diastopora patina* from the southern or lesser arctic seas have zooecia apically divided into two parts almost in the same manner, although the zooecia of *Hornera lichenoides, Idmonea atlantica*, and *Diastopora hyalina* from the northern or arctic seas all have an orifice margin that is flat, or that inclines on only one side into the form of a tooth. Just as the forms of Cyclostomata flourished greatly in older geological periods, so also did they maintain a lower grade of evolution. Consequently although this order is rich in different shapes of colonies, it exhibits zooecia that are almost the same. Nevertheless, even in regard to the zooecia of this order we can notice that the variations generally follow evolution since the zooecia become bigger and are more and more closely joined. But always in the disturbance of the law, the observations we have made above concerning the variations are valid. For example, *Idmonea atlantica* is able to retain smaller zooecia occasionally, although generally it more closely resembles *Idmonea serpens*; or *Diastopora patina* can form a large colony while retaining its zooecia spread into diagonal rows.

Moreover, concerning the Cyclostomata (*Diastopora hyalina, D. patina, Discoporella hispida, Corymbopora fungiformis*, and *Coronopora truncata*) we noted that the budding of the colony creates new layers above

lower zooecia that have been enclosed. By this budding certain genera of d'Orbigny, namely *Reptomultisparsa, Cellulipora, Semimultisparsa, Multisparsa*, etc., are to be explained. This budding of the colony indicates the same rebuilding after histolysis which we described in the case of the zooecia of the *argillacea* form of the species *Aetea truncata*.[32] Among the Cheilostomata we find this budding in the examples of *Membranipora flemingii, Escharella jacotini, E. linearis, Mollia hyalina, Myriozoum crustaceum, Porella laevis, Eschara verrucosa, E. cervicornis*, and several [species of] Celleporina. In addition this type of budding is followed in these species by a disturbance of the position and often of the form of the zooecia with the result that species of the suborder Escharina more closely resemble Celleporina.

CHEILOSTOMATA

PRINCIPLES

The Cheilostomata [table 3] reach a higher grade of evolution. This is especially apparent from the fact that the individuals of a colony have a more distinctive appearance, and those secondary organs of the colony (setae, ooecia, avicularia, vibracularia, and tendrils) are augmented. These arise very often by secondary budding on the top of zooecia although they are also able to arise from primary budding in the manner of zooecia. Furthermore, even modifications of the same type as occur in zooecia can be seen. Species maintaining a lower grade of evolution especially show this (that is those whose colonies are less differentiated). In regard to avicularia we can cite the examples of *Flustra foliacea*,[33] *Flustra securifrons*,[34] and *Flustra membranaceo-truncata*.[35] In regard to tendrils we can cite the examples of *Flustra papyrea*[36] and *Flustra securifrons*[37] in which species, however, the outer zooecia on the branches of the stem assume the role of tendrils through their extended base; compare *Bugula murrayana*,[38] and *Bicellaria alderi*.[39]

As to subdividing the order into suborders, we follow the principles that Milne-Edwards seems to have been the first to infer when, in [referring to] the notes on *Flustra* in the writings of Lamarck, Milne-Edwards said that the form of the zooecium was the only distinguishing feature of the Bryozoa which characterizes genera with a fixed delimitation. Now it appears that contemporary knowledge of these animals is better; consequently our definition of both suborders and genera is different. After Milne-Edwards, d'Orbigny gave a complete system in which he said that he followed the same principles. Actually, however, he made his divisions according to the secondary organs of the colony, whose relationships (it should be added) he did not understand. Thus he very often confused the relationships. And this is true also of the system of Gray, who added almost nothing more to our knowledge of these animals. But these authors, and Lamouroux before them, pro-

TABLE 3

CLASSIFICATION OF EXTANT SUBARCTIC AND ARCTIC CHEILOSTOME AND ENTOPROCT BRYOZOA AS GIVEN BY SMITT IN THE TABLE
IN HIS TEXT, AND MODIFIED TO AGREE WITH THE NOMENCLATURE USED IN HIS TEXT

Order 3. Cheilostomata

Suborder 1. Cellularina

Family 1. Aeteidae

Genus *Aetea*
Aetea truncata forma typica
Aetea truncata forma argillacea
Aetea anguina forma spathulata
Aetea anguina forma recta

Family 2, Cellulariidae

Genus *Eucratea*
Eucratea chelata
Genus *Cellularia*
Cellularia ternata forma ternata
Cellularia ternata forma gracilis
Cellularia ternata forma duplex
Cellularia scabra forma typica
Cellularia scabra forma elongata
Cellularia reptans
Cellularia scruposa
Cellularia peachii
Genus *Gemellaria*
Gemellaria loricata forma typica
Gemellaria loricata forma elongata
Genus *Caberea*
Caberea ellisii

Family 3. Bicellariidae

Genus *Bicellaria*
Bicellaria ciliata
Bicellaria alderi
Genus *Bugula*
Bugula avicularia
Bugula flabellata
Bugula fastigiata
Bugula murrayana
Bugula quadridentata
Bugula umbella
Genus *Beania*
Beania mirabilis

Suborder 2. Flustrina

Family 1. Flustridae

Genus *Flustra*
Flustra membranacea
Flustra membranaceo-truncata
Flustra securifrons
Flustra papyrea
Flustra foliacea

Family 2. Cellaridae

Genus *Cellaria*
Cellaria articulata (= *C. borealis*)
Cellaria fistulosa

Family 3. Membraniporidae

Genus *Membranipora*
Membranipora lineata forma craticula
Membranipora lineata forma lineata
Membranipora lineata forma discreta
Membranipora lineata forma sophiae
Membranipora lineata forma unicornis

Membranipora lineata forma americana
Membranipora nitida
Membranipora spinifera
Membranipora arctica
Membranipora flemingii forma cornigera
Membranipora flemingii forma trifolium
Membranipora flemingii forma minax
Membranipora pilosa
Membranipora laxa
Membranipora pilosa forma monostachys
Membranipora catenularia forma typica
Membranipora catenularia forma membranacea

Suborder 3. Escharina

Family 1. Eschariporidae

Genus *Escharipora*
Escharipora nitido-punctata
Escharipora punctata
Escharipora annulata

Family 2. Porinidae

Genus *Porina*
Porina malusii
Porina ciliata
Genus *Anarthropora*
Anarthropora monodon
Anarthropora borealis

Family 3. Myriozoidae

Genus *Escharella*
Subgenus *Escharella*
Escharella porifera forma typica
Escharella porifera forma minuscula
Escharella porifera forma majuscula
Escharella porifera forma edentata
Escharella porifera forma cancellata
Escharella palmata
Escharella legentilii forma prototypa
Escharella legentilii forma typica
Escharella jacontini
Subgenus *Herentia*
Escharella auriculata
Escharella landsborovii
Escharella linearis forma typica
Escharella linearis forma biaperta
Escharella linearis secundaria
Genus *Mollia*
Mollia spinifera
Mollia ansata
Mollia papillata } "*Mollia vulgaris*"
Mollia candida
Mollia hyalina (*Hippothoa* of authors)
Mollia divaricata (*Hippothoa* of authors)
Genus *Myriozoum*
Myriozoum crustaceum
Myriozoum subgracile
Myriozoum coarctatum

Family 4. Escharidae

Genus *Lepralia*
Lepralia pallasiana
Lepralia spathulifera
Lepralia hippopus

Table 3—(*Continued*)

Genus *Porella*
   Porella acutrirostris
   Porella laevis   (= *Eschara laevis*)
Genus *Eschara*
   Eschara patens
   Eschara verrucosa
   Eschara propinqua
   Eschara cervicornis type a.
   Eschara cervicornis type b.
   Eschara cervicornis type c.
   Eschara elegantula
Genus *Escharoides*
   Escharoides sarsii
   Escharoides rosacea
 	
Family 5. Discoporidae
 	
Genus *Discopora*
   Discopora scutulata
   Discopora coccinea forma peachii
   Discopora coccinea forma ventricosa
   Discopora coccinea forma ovalis
   Discopora coccinea forma labiata
   Discopora appensa
   Discopora sincera
   Discopora pavonella
   Discopora skenei
	Suborder 4. Celleporina
	
	Family 1. Celleporidae
	
	Genus *Cellepora*
	   Cellepora scabra
	   Cellepora plicata
	   Cellepora ovata
	   Cellepora ramulosa forma contigua
	   Cellepora ramulosa forma ramulosa
	   Cellepora ramulosa forma avicularis
	   Cellepora ramulosa forma tuberosa
	Genus *Celleporaria*
	   Celleporaria hassallii
	   Celleporaria incrassata
	
	Family 2. Reteporidae
	
	Genus *Retepora*
	   Retepora beaniana
	   Retepora cellulosa
	   Retepora notopachys var. elongata
	
	Tribe 2. Hippocrepia
	
	Family 1. Pedicellinidae
	
	Genus *Pedicellina*
	   Pedicellina echinata
	   Pedicellina gracilis

posed so many names of genera that the most recently named suborders in particular abound with these. In order that the number does not now grow too large, we have retained those genera, although very often it was necessary for us to change their meaning. Busk gave the best recognition of forms, although he himself admitted that he used an artificial system. Thus, so that our enumeration may be understood more easily, we also add in parentheses the genera of Busk. [These names, generally omitted in the remainder of Smitt's discussion, are not included in our rendition of tables 1–3.]

We wish first to observe that there are almost no traits so fixed that they do not vary in the evolution of forms. Because of this we shall try in vain to describe with definite traits the limits of orders, families, and genera (often also of species). These series, having evolved from the same origin, resemble each other in many ways. Often the forms of individual series are so similar that it is very difficult to decide where a given form is to be referred. Therefore, we are not able to do anything other than follow through the evolution of each form so that we may have a definite decision about its nature. We have observed the evolution of most of our species. According to this—and by analogy in accordance with the form of zooecium, where we have not been able to follow the evolution—if they continue the same or similar road of evolution, our practice is to arrange them into suborders, families, and genera. But when a given species has been described in one stage, or in a less-evolved stage, a well-known, major difficulty arises, namely of assessing its true relation-

ship. A great uncertainty in the synonyms results from this, particularly concerning extinct species. Nevertheless it is very important to know them so that they may fill the gaps which occur among living species. Furthermore d'Orbigny believed rightly that the families are able to show many modes of colony construction. But so that we do not distribute species into genera according to this principle, let us just observe that many species exist which are first in the form of an incrustation and then have been built up to the form of a stem, or which are first connected and then have been broken up into the form of *Hippothoa*. Often these variations of colonies cause other corresponding variations which influence the form of zooecia. For instance *Eucratea* of the family Cellularidae, or *Beania* of the family Bicellariidae imitates the form of *Aetea,* or *Caberea* of that first family, or *Bugula* of the second, when they form additional series of zooecia, display them almost like the Flustrina. Or *Membranipora catenularia* and *Mollia divaricata* (*Hippothoa* of authors) are built up from the zooecia that are almost cellularine, or *Catenicella* from the suborder Escharina offers the same similarity to the Cellularina. Or those forms of escharine animals noted above, which make new strata of zooecia on top of the lower ones, very often make celleporinelike zooecia. Let these be examples for us so that we may determine species only after the evolution of forms has been thoroughly understood so that we do not consider differences of this type to be definite traits, even though the type of zooecium may *seem* different. Nevertheless, it will be idle for us to indicate other traits or principles for the classification of Bryozoa for we shall always see that wherever the form of the zooecium is the same, the other dissimilarities vanish in one way or another.

Now that we have given these remarks, we recognize four suborders of Cheilostomata: [1] Flustrina, whose square zooecia have a flat frontal which is equal to the primary region of the opesium. Therefore they commonly maintain the lowest grade of evolution in that the Cheilostomata evolve from stages corresponding to the Ctenostomata in just such a way that they construct a proximal part behind the region of the opesium and restrict that region more and more; [2] Cellularina, whose horny or horny-calcareous zooecia are funnel-shaped and exhibit a proximal (lower) part that is tubular or rather conical below the region of the opesium; [3] Escharina, whose generally calcareous, square or semi-oval zooecia extend recumbently in the plane of the enlarging colony (which is either encrusting or raised into the shape of a stem), and which exhibit a lateral opesium whose size generally equals the size of the operculum, with no additional continuous frontal area of the zooecium remaining; [4] Celleporina, whose calcareous, rhomboid, and oval zooecia, having been more or less built up to the plane of the enlarging colony, and unevenly heaped up, have a terminal orifice.

CELLULARINA

First let us consider the Cellularina whose relationship, expressed through *Aetea*, with *Vesicularia* and the Cyclostomata we have already noted. The tubular-shaped zooecia of the Aeteidae [40] are exposed at the apex in the manner of the Cyclostomata. Nevertheless they have the region of the opesium on the side, a characteristic of the Cheilostomata. The tentacular sheath of the individuals is strengthened by a ring of spines in the manner of the Ctenostomata. By these traits the Aeteidae join all three orders of funnel-like Bryozoa. In their variations, however, they especially resemble *Vesicularia*. That is, zooecia are lengthened more and more in proportion to the evolution of the forms. And the colony, at first creeping, is then built up [41] to the form of a stem [42] and is often composed of verticillate zooecia. [43]

The Cellulariidae include forms from the two suborders of d'Orbigny (who nevertheless gave up this way of classification afterwards) and of Busk, namely the articulata and the inarticulata. The similarity of the zooecia, and the positioning of the secondary organs of the colony, which is the same, show that these should be joined. Both *Eucratea* and *Gemellaria* have the zooecia that are lower on the branches constricted proximally, whereas other genera have an articulation. *Caberea* retains branches that are separated at the base but that are joined together by tendrils, in the same way in which *Escharella palmata*, a species of the suborder Escharina, compensates for articulations. If we consider the form of the secondary organs of the colony,

our fauna even shows in the case of *Cellularia scabra* and *Cellularia reptans,* a transition to those large vibracularia of *Caberea.*

Species of the genus *Cellularia* illustrate the evolution of the colony proceeding from a first zooecium, which is more or less a tata-form [44] (Genus *Tata,* v. Ben.). There grows from this a second zooecium that is already very nearly typical, and whose region of the opesium lacks proximal spines along the margin. The shape of the well-known tata-form zooecium of individual species differs in such a way that this [second] zooecium is more or less [typical] of the Cellularina. Often the species themselves show the same differences; [45] thus we do not always see them proceeding from the same stage, just as we already noted above that the early stages seem by no means to be always necessary. The evolved forms show differences that have arisen according to the laws of evolution. Thus that species, for which we have retained the name *Cellularia ternata,* offers a progressive sequence of forms [46] as we have shown in the cases of *Crisia eburnea* and of *Crisia denticulata.*

The family Bicellariidae has colonies which commonly grow less rigidly so that their parts are more flexible. It is distinguished by twisted zooecia, whose region of the opesium is displaced laterally and obliquely relative to the median plane of the zooecial axis; thus the lower corner of the frontal area is closer to the internal than to the external side of the zooecium. Very often we see this torsion illustrated in a conical colony whose shape is determined by the spiral pattern of its branches.

We have also been able to pursue thoroughly the evolution of the species of this family. [47] The first zooecia have spines all around the margin of the region of the opesium, although later zooecia commonly retain spines only on the distal side of this region. Later-formed zooecia diverge farther from the shape of the tata, when their proximal part has been extended and they become erect. It is especially necessary to consider this pattern of evolution if we wish to distinguish species of *Bugula.* Writers on this topic have distinguished species especially by the number of spines. Thus we arrange *Bugula avicularia, B. flabellata, B. fastigiata,* and also *B. murrayana* and *B. quadridentata* into series, whose members show that, in the manner of *Crisia,* they pass their more or less constant stages of evolution as a separate existence. *Bugula umbella* [48] and *Beania mirabilis* show a large deviation from the typical form of zooecia. But of these the former has twisted zooecia and avicularia typical of *Bugula,* whereas the latter combines an arrangement of its spines, which is similar to *Bugula murrayana,* with a similarity to the zooecia of *Bugula umbella.*

FLUSTRINA

Already above we noted the manner in which the [suborder] Flustrina approaches the Ctenostomata.

This is especially valid for the family Flustridae. Their forms in our fauna have a stage of differentiation slightly above the Ctenostomata. Concerning the form of the colony, which is considered a generic trait by many authors on this matter, the species of *Flustra* offer the best examples to us to show how insignificant this is. Both the simple ones and those formed of two layers of zooecia grow encrusting and erect into the form of a lamina that is undulating, or unequally or labially foliate. The form of the zooecium of the species maintaining a lower grade of differentiation is quite general although according to it we do make two subdivisions [within the suborder Flustrina.]

The external rectangular form of the zooecium of *Cellaria borealis* [49] shows that the family Cellariidae belongs with the suborder Flustrina. Besides, no one will doubt this who has seen the differences in the evolution of the zooecia in *Cellaria fistulosa*. For instance the orifice is transferred almost to the middle of the frontal of the zooecium by the growing of this distal part,[50] whereas the young zooecia resemble the form of the zooecia of *Flustra foliacea*. This change seems to be of great importance for revealing the structure of *Melicerita*, which may therefore be shown to be a form of this family, although we see an entirely different mode of construction of the colony in that genus. And I suspect that many forms ought to be transferred from the genera *Vincularina* and *Escharinella* d'Orbigny and *Micropora* Gray into this family; however, I have not discerned the [pattern of] evolution, nor can a definite judgment be made from the illustrations of others.

This is also valid in regard to the Membraniporidae. For example, we are by no means able to judge with certainty the relationship between *Membranipora arctica* and *Biflustra*, although there is a great affinity.

We have been able to follow four species of the family Membraniporidae, *Membranipora lineata*,[51] *M. flemingii*,[52] *M. arctica*, and *M. pilosa*,[53] that start with the true form of the tata. They undergo in their evolution many variations which seem to represent distinct species, and by which they have many interconnections. Deriving our information from other forms and from the precise nature of their evolution, we assign those variations to series within the limits of species. Very often a great dissimilarity results from an increase in calcification or from a decrease from that harder construction, which we frequently see in stagnant or less salty water where these forms are reduced to the lowest stages of calcification (*Flustra membranacea*,[54] Müll., *Millepora reticulum*, Lam.?, *Membranipora lacroixii*, Aud., Busk?, *Membr. hexagona*, Id.?). But if we examine the traits by which these varieties are distinguished in the manner of species by scholars, we will recognize that they are all indications of stages of evolution, or [to put it another way] that they are related by intermediate forms. It would take me too far if I decided to describe these relationships again. Let it be enough

for me to note the decreasing number of spines,[55] the progressive restriction of the region of the opesium by the proximal calcareous lamina,[56] the increasing hardness and pointedness of the ooecia (which is very frequently made more prominent by a diagonal calcareous ridge extending above the ooecium and pointed in the middle [57]), and the progressive number and size of avicularia which are the organs of defense of the colony and which can also vary in position (*Membranipora trifolium* [58]). After these remarks our survey will easily show the series of forms and their interrelations. We have added *Membranipora nitida* to this family because of the whole structure of the zooecium and of the colony (in the form of *Mollia*, of authors [59]), which is closely related to *Membranipora lineata*. But just as we shall see the forms of the [suborder] Escharina proceeding from tati-form origins, where they resemble *Membranipora*, so also *Membranipora nitida* will be of great importance for revealing the structure of certain forms from that suborder. We have proposed as a new form *Membranipora laxa* [60] from the series of *Membranipora pilosa* since it shows a colony growing like *Eucratea*. In the Arctic Ocean we see *Membranipora catenularia* [61] that is very often sheetlike and grows with the typical form of *Membranipora*; it has a form like *Hippothoa* in the abyssal regions of the Atlantic Ocean, whereas in the Oeresund Strait it is found sheetlike.[62] From there, as the reduction of its structure proceeds more and more,[63] we have been able to follow it as far as Arkö (East Gotland) since it has been left behind in the inner Baltic Sea from the glacial era (see Lovén). But there occurs here what is often found when several series of forms proceed from a similar origin, namely that they may look like each other in many ways. Consequently, in arctic waters we see a form of *Membranipora lineata* [64] that is elongated and very hard because of calcification. Perhaps with equal propriety, we ought to join [*M.*] *catenularia* to this although the form of the zooecium and the big proximal spine at the region of the opesium are characteristic of *Membranipora pilosa*.

ESCHARINA

There has been great confusion concerning [the suborder] Escharina. Authors have established genera according to the form of the colony, or the characteristics of the zooecium, whose significance was unknown or nil. Hence, related species have been too far removed from each other. Because the literature contains no observations about the evolution of the forms, we must make almost a whole new arrangement for these. For although we are able to find scattered observations in other authors concerning the differences of forms, nevertheless they have been of almost no help in [understanding] a natural system, until we reach Hassall, who thoroughly examined the variations of the genus *Lepralia* (Johnst.) and found a very constant shape of orifice. This observation will very often show us also the path to a natural system, once we reject the opinion of earlier authors concerning the formation of genera and families; but we shall see that even this characteristic is subjected to variation, and therefore we shall retain the mode of variation as the sole principle for the classification of forms.

To turn to our first family, the Eschariporidae (whose name we borrow from d'Orbigny), this includes forms which are not typical of our fauna but which clearly indicate a connection with *Membranipora*. This family especially flourished during the time of formation of Cretaceous [rocks] from which d'Orbigny described the erect forms of the colony called *Escharipora* (e.g., *E. filiformis*) and those well-known [species of] *Steginopora*. At present, we find that only the encrusting species are flourishing. Of these, *Escharipora nitido-punctata*, an arctic form of the well-known *E. figularis* which is equipped with lateral avicularia at the zooecial orfice,[65] shows the front of the zooecium constructed from struts that come together to form a dome and corresponding to the marginal spines (costae) of *Membranipora nitida*. The fissures between the struts, which are comparable to the pores of the zooecium of *E. punctata* [66] and of *E. annulata*,[67] represent just one type of evolution of the morphology of [the suborder] Escharina. In addition, the orifice of the zooecium [68] of this family is transversely elliptical, more or less elevated, and sometimes somewhat quadrangular; this [characteristic] indicates also the affinity of the family Discoporidae which is much closer to [the suborder] Celleporina. We shall see that *Discopora scutulata* [69] forms its zooecia in almost the same way [as the Eschariporidae]. But one of its pores, almost in the middle in the front of the zooecium, is crescent-shaped. In the case of *Escharipora innominata* this already shows an easier transition to the structure of the family whose best characteristic for distinguishing it also formed the basis for d'Orbigny's family Porinidae.

But what is to be considered the outstanding feature of this family [Porinidae] is the semicircular form of the zooecial orifice which is straight and unbroken at the proximal margin. When both this characteristic and that pore are able to be elongated into a tubular shape, the one characteristic may help the other for identification. The physiological significance of this pore is still unknown, but *Anarthropora monodon* shows a transformation [of the pore] to an avicularium. Thus we are able to postulate a morphological analogy between it [the pore] and that common medial avicularium of other [species of the suborder] Escharina, which is proximal to the zooecial orifice.

We include the especially typical forms of this family in the genus *Porina*, whose species *ciliata* occurs erect in the manner of *Eschara* of authors [70] in the Mediterranean Sea (*fide* Copenhagen Museum). Other species under the name of *Tubucellaria* (d'Orb.; *Onchopora*,

Bsk.) have been already known for a long time, of which one (*Tubucellaria mutica*, Bsk.) shows a transition of the form of the zooecium to *Anarthropora*. This is encrusting or built up into a stalk lacking points and is distinguished by the elongation of the zooecial orifice and of the medial pore into the form of a little tube. But *Anarthropora borealis* [71] by its elongate zooecia and its mode of calcification, is very similar to the Pustuliporidae among the Cyclostomata just as another genus of the Cheilostomata, namely *Retepora*, shows the same similarity to the Horneridae. As a consequence, these two genera formerly were placed among the Cyclostomata. This happens as the evolution of the type progresses when [the suborder] Escharina approaches more and more to the [suborder] Celleporina, whose terminal orifice of the zooecium more easily produces variations similar to the Cyclostomata.

A third family of [the suborder] Escharina, to which we have given the name Myriozoidae, contains several series of evolution, which we designate by the names of the genera and subgenera. We can investigate its evolution in two ways. One [mode of evolution] is the true evolution of the colony from the first zooecium, and the other [mode of] evolution is of the zooecia from a common bud. The former we have pursued in regard to *Escharella porifera*, where the first zooecium has the form of *Membranipora*. The region of the opesium is elliptical and is covered over at its most proximal portion by a calcareous concave lamina which lacks the pores of the typical zooecium of this species. Now this zooecium is closed more quickly and is covered with an advancing calcification so that its primary form is not able to be distinguished.[72] More often we have observed the other [mode of] evolution, which shows differences of the zooecia from the common bud up to the adult stage. This begins very often with stages that are almost Flustriform when we see the frontal plane of the zooecium and its rectangular or hexagonal shape (i.e. a truncated rhombus). Furthermore, the area of the opesium of the zooecium has a circular-square or a circular form. More and more the younger zooecia of the colony, or those that are more highly evolved, form a convex frontal; (the zooecia however, are able to be flattened with increasing calcification). In the region of the orifice of the zooecium at the proximal corners calcareous laminae build up [73] which, by the appearance of lateral processes, fill in the space in front of the articulation of the operculum, until they approach [each other] in the middle of the proximal margin, so that they leave only a sinus in the middle,[74] or so that the proximal margin of the orifice grows into a medial tooth.[75] The forms, furthermore, which are less highly evolved, as in the preceding families, can be distinguished by the frontal portion of the zooecia that is pierced with larger primary pores;[76] but when evolution has further advanced, these pores are filled, or secondary pores grow above an almost continuous skeleton, in such a way that a calcareous line is raised at the boundaries of the zooecia, from which other lines may diverge on top of the zooecial frontal which afterwards are interconnected in the form of a network and may more and more conceal the zooecia.[77]

The first genus, *Escharella*, shows the greatest number of these variations. They cause very great differences in the growth form of *Escharella porifera*, which is in a way the prototype form of other species, although we are able to follow the development even of these from the lowest stages of the structure of the zooecium. Here a good norm, if we want to estimate the grade of differentiation, is offered to us by the form of the mandible of the median avicularium. This form is semielliptical in *Escharella porifera*,[78] and varies between this and the shape of a triangle in *Escharella palmata*,[79] and is triangular in *Escharella legentilii*.[80] We have previously noted the significance of these differences in *Membranipora* in which the pointed form of the mandible (i.e., the operculum) of the avicularium indicates a higher grade of evolution. *Escharella palmata* shows a quite strange method of growth; its stem, which is erect, flattened, and dichotomously branched, is transversely subdivided and then is held together by horny tubules.[81] We have already said that this is what ought to be noticed in *Caberea ellisii* although in this species of *Escharella* it is easier to investigate. In the prototype arctic form of *Escharella legentilii*,[82] we see that the zooecia lack the secondary margin of the orifice, and that the medial avicularium placed traversely here and there diverges from the apex. Now the typical form of this species, such as *Escharella palmata*, builds up an S-shaped secondary margin of the zooecial orifice.[83] This edge often merges above the front of the ooecium from each side, and creates a secondary orifice, whose lateral margins merge with the edges of the orifice of the avicularium. Thus those three orifices (of the ooecium, zooecium, and avicularium) are joined. This form of secondary orifice also occurs in *Escharella jacotini*, although this species lacks the proximal medial avicularium at the zooecial orifice. The significance of this species for classification is best expressed by the fact that it shows a transition from those preceding to the conformation of the [suborder] Celleporina. For it shows almost the same form of zooecium and of its orifice, and the same mode of calcification[84] as in *Escharella legentilii*. In this species the avicularia are lateral, more variable, and more built up, and distributed unevenly on top of most, but not all zooecia. And the zooecia of the younger strata,[85] which the colonies build out over the older zooecia, are unevenly heaped up and more inflated and very often show a deformed orifice.

Herentia, a subgenus of this genus, shows almost the same differences in its evolution of zooecia. Yet these differences proceed in another way with the zooecia very often possessing an orifice standing more

by itself, and bent proximally in the middle. Since the form of these [zooecia] is flattened also in these species, they [the species] are distinguished more easily from those preceding by the mandible of the medial avicularium, which being rather short very often resembles a semicircle.[86] And the forms have almost the same interrelationships, so that we may consider *Herentia auriculata*,[87] which leads its own existence in the manner of a species, as a prototype form in a certain sense of *Herentia landsborovii*[88] although it approaches nearer and nearer to the less-evolved stages of *Herentia landsborovii*. *Herentia linearis*[89] shows a transition to the conformation of the [suborder] Celleporina in the same way as *Escharella jacotini*. Now of its [*H.*] variations, we further wish to note the spines that are commonly pressed close together and are bigger than those which arise at the margin of the region of the opesium of the zooecium. By an uneven calcification of the skeleton, they [spines] are warty like zooecia. And (one or two) [spines] occupy the place of the median avicularium of other species and can arise in place of a lateral avicularium of this very species. The avicularia (?) which are rather unusual in this species[90] have the form of ooecia, as we have seen, and are irregularly scattered over the colony. They especially recall the "*cellulae fertiles*" which Busk described. Nevertheless, we cannot believe that that fancied interpretation regarding the *cellulae fertiles* is correct since the operculum indicates the action of an avicularium, by its own strong pressure of muscles. Therefore, we rather would think these ought to be compared with the large, commonly-occurring avicularia of the suborder Celleporina. Finally, we see two arctic forms[91] that are so closely connected with *Herentia linearis*, that we include them within the limits of that species. Yet, these are especially remarkable because they have a shorter zooecial orifice with a semicircular proximal edge that is curved in the middle, and more and more straight on the sides. And so they show a transition to *Mollia*.

The genus *Mollia* often shows many differences of evolution and such constant variations that it is considered correct to distinguish species. If we recall the known laws of evolution, the differences of calcification, and the variations of the secondary organs of the colony that we have already often mentioned, then these forms will easily be able to be arranged into a series of species. Thus we see *Mollia spinifera*[92] as a flattened, porous form and equipped with very many spines at the edge of the zooecial orifice. As to *Mollia ansata*, although we have been able to follow it from the lowest stages of evolution,[93] we think that it is nothing other than a form[94] of the same series which is somewhat elongated, commonly less porous, and lacking the spines. Often we find this form pointed (the mucro (point) arises from the transformation of the medial proximal spine at the zooecial orifice of other forms); however, this variation is better expressed by *Mollia papillata* of which even the transverse calcareous lines of the zooecium we can see already indicated in the preceding form. Finally *Mollia candida*[95] shows the shortest form of zooecial orifice, to which the other members of this series more or less approach; in addition to this, having advanced less in calcification, it maintains a more uniform appearance, lacking ooecia and avicularia, as far as is known for it [*M. candida*]. Thus we say that it is a form maintaining the highest stage in the variation of the orifice but lower stages in other respects, just as we often have noted a greater constancy in the evolution of a certain difference in one form than in another in regard to species of other series.

Furthermore, it appears that *Mollia vulgaris* shows a great analogy to *Eschcaripora ciliata*, whose [*M. v.*] proximal sinus of the zooecial orifice corresponds to its [*E. c.*] medial pore. If we inspect extinct and peculiar forms, we even see analogous variations. Furthermore, the series *Escharipora granifera-Escharipora pyriformis* shows the same relation to *Escharipora ciliata* as the variations of *Mollia hyalina*[96] show to *Mollia vulgaris*; in addition, the Celleporines suggest a similar development even more. *Mollia divaricata*,[97] although it seems to be a separate species, is nothing other than a variation of the form of *Hippothoa*, which we can already see in the case of *Mollia ansata*,[98] except that it has larger zooecia.[99]

The genus *Myriozoum* shows the same differentiation of form of the zooecial orifice if we compare the [*M.*] *truncatum* form of the Mediterranean Sea with our forms, which we noted above as an example of the manner of evolution, and which at least one arctic form, *Myriozoum coarctatum*, showed us among the stages of its growth. Therefore, according to this order of evolution, *Myriozoum trunctatum* maintains a lower grade of evolution. The [zooecial] form of *Myriozoum crustaceum* is very similar to the zooecial form of *Mollia spinifera*. *Myriozoum crustaceum*, however, when the calcification is more advanced,[100] is so similar to *Eschara incisa*, which Milne-Edwards described, that I do not think that these forms ought to be distinguished. But it happens, although we still are not able fully to explain it, that we see absolutely the same conformation and relative position of zooecia and avicularia in *Myriozoum crustaceum* as in *Myriozoum subgracile*. The only noteworthy difference comes from conformation of the colony. Although the former never seems to exist with the latter, we have very often seen it [*M. c.*] as an incrustation on top of the branches of *Myriozoum coarctatum*. The upright form of the *Myriozoum* colony, out of which Sars wanted to form a new genus, is especially noteworthy for the reason that it apparently must be explained through a comparison of the articullate forms. It forms constrictions in two ways. The zooecia are either enclosed in a transverse ring (which rather recalls the joints of *Caberea ellisii* and *Escharella palmata*), where afterwards the enlargement of the stem

ceases (or perhaps it is constricted, but we have been unable to see this), or the zooecia are elongated in a transverse ring, in the way in which the so-called articulata make their own articulation. And in truth *Cellaria*, when it has calcified articulations, has a great similarity to *Myriozoum*.

The family Escharidae follows another path in the evolution of differentiations; their zooecial orifice is reshaped from a form that is commonly semi-elliptical,[101] such as *Escharipora malusii* shows, first to a form that is proximally constricted on the sides (at the hinges of the operculum) (claviform[102]) and is either semicircular,[103] or circular.[104] The secondary orifice, which is claviform where it projects from the elevation of the edges, is bent proximally in front of the immersed medial avicularium,[105] which we have already observed in preceding families.

Concerning three species of this family, *Lepralia spathulifera*,[106] *Lepralia hippopus*, and *Eschara laevis*,[107] we have noticed progressive differences in the colonies of zooecia from the form of the tata. First,[108] the proximal part [of the zooecium] and the region of the opesium are elongated. Then[109] the proximal part of the region of the opesium is covered, as in the Membraniporidae, by a calcareous lamina. Next[110] the proximal spines occur at the edge of this region. Then[111] the proximal part of the margin of the region is opened in such a way that the skeleton of the proximal part of the zooecium and the lamina of its frontal areas become continuous. Eventually the whole proximal part of the zooecium becomes equally convex behind the orifice,[112] to which we now give this name, so that it is distinguished from the primary frontal area, since the proximal vestiges of this margin are commonly destroyed. In the evolutionary differences of *Membranipora flemingii* we have already seen the prototype of this evolution. We have also noted another type of relationship between the [suborder] Escharina and the [suborder] Flustrina in which the flattened spines (costae) of *Membranipora nitida* can be recognized in the formation of the front of the zooecium of Escharina. According to these two ways of evolution, Escharina perhaps will at some time be able to be distributed into two series, but at present the observations are insufficient. Therefore, let it be enough for us to say that its origin from the Flustrina type is certain.

Now it happens even here that we see a rather great unevenness in degree of evolution, whereby the first zooecium of one colony[113] is tata-form, but of another colony[114] this zooecium has the form of the same species of *Membranipora* that is more evolved, but in yet another colony this zooecium shows a transition[115] to the form of Escharina. We have also observed this in colonies of the same species that are attached to the same shell. Thus we think it is again worth mentioning that a particular species is able to begin its own evolution from a higher or lower stage of development.

We divide the family Escharidae into four genera, for one of which we keep the name *Lepralia*;[116] it retains almost a uniform type of escharine appearance, and can generally be distinguished easily by the concavely curved proximal edge of the orifice. The second genus, to which we give the name *Porella*,[117] to use an old name in a special way, has its [orificial] margin convexly curved, and it has the proximal part of the zooecial orifice around a medial avicularium to form a half ring.[118] In the third genus, for which we keep the name *Eschara*,[119] a medial avicularium at the orifice of the zooecium is proximally elongated,[120] or is built up to the form of a beak[121] and leaves the primary margin of this orifice free over its greater part. Already certain forms in this genus advance farther toward the conformation of the [suborder] Celleporina, for *Eschara patens*[122] and *Eschara propinqua*[123] show larger lateral avicularia. And this form builds up lateral margins, which are secondarily S-shaped, of the zooecial orifice; in these margins smaller internal avicularia (those lateral to the zooecial orifice) sometimes arise.[124] And in the case of *Eschara cervicornis* the difference between the external and internal layers of the zooecia, which shows a transition to the [suborder] Celleporina, has been already known for a long time. Furthermore, concerning this species one must note the difference in the geographical distribution of forms. For there lives in Greenland a form[125] that is quite frequently with an encrusting appearance (*Lepralia* of authors), which Goes collected, and is raised up into the form of *Hemeschara* at the Spitsbergen Islands. The form of *Eschara* which we find in the arctic regions commonly to be broader because of the branches or laminae of the colony, becomes slender in the south in such a way that the person who has not seen the intermediate forms regards the form from the Mediterranean Sea as another species. We are able to note this difference of the form of the colony also in *Eschara elegantula*, which we have found to be both layered and branching[126] in the Arctic Sea; it is more slender in the North Sea. The fourth genus of this family, which can be called *Escharoides*, shows in its own way a transition to the conformation of the [suborder] Celleporina. For it retains almost the same form of zooecium as the preceding genera have shown, in such a way that *Escharoides sarsii*[127] seems to be very close to *Eschara cervicornis*, and *Escharoides rosacea*[128] seems to be very close to *Porella laevis*, but these [E. *sarsii* and E. *rosacea*] have an avicularium,[129] or two avicularia,[130] lateral to the zooecial orifice.

The family Discoproidae, whose forms are commonly very much calcified, changes its zooecial orifice from a round or semielliptical form[131] to a form that is quadrangular and commonly longer transversely.[132] Very often it has a prominent tooth in the middle of the proximal margin of the orifice.[133] The species which we mention first, *Discopora scutulata*,[134] is to be referred to this family because of the form of its zooecial orifice; nevertheless because of the frontal area of the zooecium which is perforated with pores, it resembles the structure of the Eschariporidae. Thus the only thing we can say for certain is that a relatively lower stage of evolution is occupied by this species, which should be referred to this family because of a trait that is usually of very great importance. For we have observed another method of evolution in the case of *Discopora coccinea*,[135] whose colony we have found to be evolved[136] from a first zooecium that is very similar to *Membranipora flemingii*. But if we try to distinguish the forms of *Discopora coccinea* as distinct species, then we shall see the same relations among these as those which have already often been mentioned between *Crisia* and others.

CELLEPORINA

The [suborder] Escharina has often already approached nearer and nearer to the structure of the [suborder] Celleporina. And among the Celleporina, the first forms which we enumerate[137] show almost the same construction of the zooecium as *Eschara verrucosa*. They are distinguished from this species particularly by the avicularium which is oblique to the proximal edge of the zooecial orifice. Larger lateral avicularia[138] also occur which often obscure the whole frontal of the zooecium behind the orifice so that when the ooecia are also developed, nothing else can be seen except these and the avicularia between which the zooecial orifice is immersed in the manner already described by Fabricius in *Cellepora scabra*. The elevated interstitial lines[139] also cause a great difference in appearance; at the edges of the zooecia and radially converging from these, they grow over the sloping front of the zooecium with progressive calcification, as we see especially in *Cellepora plicata*. We have described an elongated form *Cellepora ovata*[140] of the same type, which nevertheless we have found as a rare thing only in the Arctic Sea. The structure of the colony becomes more and more uneven when it builds up many layers of zooecia. And in addition the differences which we have noted in *Cellepora scabra* are also found together in *Cellepora ramulosa*, so that, with the increasing size of the lateral avicularia[141] and with the elongated form of the zooecium[142] and of the avicularium, which is raised into the form of a beak at the orifice,[143] we would think that more species ought to be distinguished, if we did not know about the intermediate forms and the differences in their evolution. But these forms often occur with their own geographical and bathymetrical distribution, so far as we know; this, however, is not startling to find if we consider relationships already known from other forms. And in the case of the form of the avicularium of this species sometimes smaller lateral avicularia are elongated into the form of a beak and are built up at the margin of the zooecial orifice.[144] We find these also just as we do the larger lateral avicularia in the genus *Celleporaria*,[145] where, however, the proximal avicularium is lacking at the medial margin [of the orifice]. Futhermore it is apparent that, since the forms of this family are difficult to distinguish, their synonyms must be cited most cautiously. It is possible that forms in distant oceans have stages of evolution that are corresponding or very similar, having been derived from very dissimilar origins. However, in regard to *Celleporaria incrassata*, which d'Orbigny had also collected from the Mediterranean Sea, we think, after looking at species from that sea, that it ought to be joined with the arctic form.

This admonition to be careful should be even stronger in the family Reteporidae, whose arctic-boreal and Mediterranean forms seem to form one series, although specimens from those seas are somewhat dissimilar. For the typical *Retepora beaniana*[146] is distinguished by a proximal median avicularium that has a beak and is at the zooecial orifice; this avicularium is reduced to the form of a small cleft[147] or a little tube,[148] or often is absent in *Retepora cellulosa* and *Retepora elongata*, in which larger avicularia[149] arise scattered on top of the front of the zooecia and on the dorsal part of the colony. Now *Retepora beaniana* has a Mediterranean variety whose sessile median avicularium is oblique to the zooecial orifice, in a way such that it shows a transition to *R. cellulosa* (s. str.). And *Retepora notopachys*, which we consider to be typical, in the Mediterranean Sea has ooecia which always have clefts. But this form is distinguished from the other two especially by the secondary margin of the zooecial orifice, which is built up on each side in an *S*-form from the medial sinus so that it has a double sinus.[150]

SUMMARY

So then, we have distinguished and reviewed the families, genera, and species of the Cheilostomata according to the laws of science, but again and again we must remind ourselves of our ignorance, especially of extinct species, which makes the limits of these divisions unsure. We can say for certain that these forms, whose colonies and individuals can arrest their evolution in one or another stage, have progressed from a similar origin. It is easy to distinguish these stages as species. But by variations in calcification they often become so similar that we look in vain for specific characteristics. Therefore the current system is interrupted and uncertain in many places. But we have seen the differences of evolution, whose causes (whether external or internal in such a way that a more evolved stage produces other

stages of higher evolution) we still do not know, form series. These series are parallel with series of differences among species, genera, and divisions of a higher order, and clarify these in such a way that they may be referred to a similar origin. Or, if we may use theoretical language: just as individuals, so also species and the other divisions, which make up the system of nature, are subject to the laws of evolution.

ENTOPROCTA

Concerning the relation of the Hippocrepia [Entoprocta] to the other Bryozoa we still have no sure observations, but nevertheless we are not able to believe that the manner of explanation, which Allman has given, is valid. For we see the same structure in *Pedicellina* as in the Infundibulata [Ectoprocta,] by noticing only the ovary which is enclosed within a special membrane and which fills almost the whole perigastric cavity of the body, and the ring of tentacles open on one side, which, however, does not cause the analogy to be destroyed in any way; therefore we think that the tentacles of both of these should be assessed by direct comparison. But if we may suspect a grouping of forms, then it seems that these Bryozoa also ought to be closely compared to the Ctenostomata. The Pedicellinidae also show the same varieties of colony growth as does *Vesicularia*. Therefore, we think that *Pedicellina belgica* (v. Ben.) and *Pedicellina nutans* (Dalyell) are nothing other than varieties of *Pedicellina gracilis* (Sars), of which we have seen forms corresponding to the colony of *Vesicularia*.

ADDENDA

When this paper had already been handed over to the printer, I made certain new observations about these animals.

From the Mediterranean Sea I was able to study the form of *Escharipora radiata* more carefully, which because of the median crescent-shape pore of the zooecia seems to connect this genus with *Porina* so closely that the family Eschariporidae embraces the family which we have made of the Porinidae.

Retepora beaniana, in our regions, has a minuscule lateral avicularium at the zooecial orifice, although it is rare, closed by a triangular mandible, so that the transition to *R. cellulosa* is now more evident.

Celleporaria incrassata shows the evolution of a colony from the very first tata-form zooecium. But another colony has its zooecium built up in a manner like a celleporine. Compare above in regard to *Lepralia*, p. 26.

APPENDIX

Figures in *Öfversigt af Kongl. Vetenskaps-Akademiens Förhandlingar* (1865–1868) to which Smitt refers in "Bryozoa Marina." Numbers in bold-face type correspond to superscript numbers in the text. -

1. 1865*b*: pl. V, fig. 9, with pl. I, fig. 1.
2. 1867: pl. XII, fig. 6. **3.** pl. XII, figs. 15, 16, 20, 21. **4.** figs. 22–24. **5.** pl. XIII, figs. 36, 38, 39.
6. 1865*a*: p. 122, fig. 2. **7.** pl. XVI.
8. 1867: pl. VIII, figs. 1–6. **9.** fig. 10. **10.** pl. III, figs. 1, 2, 4, 6; pl. V; pl. IX, figs. 4–9; pl. X, fig. 2. **11.** pl. VI, fig. 2. **12.** pl. X, figs. 6, 7. **13.** compare also pl. XI, fig. 6. **14.** pl. XI, fig. 13. **15.** pl. VIII, fig. 6. **16.** pl. IX, figs. D1, D2; pl. X, figs. A, B. C. **17.** pl. III, fig. 6. **18.** pl. IV, figs. 4–9. **19.** pl. IV, fig. 10. **20.** compare pl. III, figs. 4, 5, with pl. IV, fig. 5. **21.** pl. III, figs. 1–3. **22.** pl. IX, figs. 2*a*, 2*b*. **23.** pl. VII, figs. 7–10. **24.** fig. 5. **25.** pl. VI, fig. 2. **26.** fig. 4. **27.** pl. XI, figs. 1–4.
28. 1865*b*: pl. I, figs. 1, 12, pl. IV, fig. 16; 1876, pl. VIII, figs. 9, 10, 13, pl. X, figs. 6, 7, C, pl. XI, figs. 1–5, 7–12.
29. 1865*b*: pl. I, figs. 13–18, pl. II, fig. 2; Öfvers., 1867: pl. IV, fig. 8; 1868*a*: pl. XVI, fig. 24.
30. Compare 1867: pl. III, fig. 2, with pl. IX, fig. 9. **31.** pl. III, figs. 3*a*, 3*b*.
32. 1865*b*: pl. IV, fig. 18. **33.** 1868*a*: pl. XX, figs. 12–14. **34.** figs. 7, 8. **35.** figs. 2–4. **36.** fig. 10. **37.** fig. 7. **38.** pl. XVIII, figs. 20, 23, 26. **39.** figs., 5, 6. **40.** 1865*b*, pl. III, figs. 3–8; 1868a, pl. XVI, figs. 3, 5.
41. 1865*b*, pl. III, fig. 1. **42.** pl. II, fig. 14. **43.** pl. III, fig. 2.
44. 1868*a*, pl. XVI, fig. 15; pl. XVII, figs. 27, 28, 42–47. **45.** compare pl. XVII, figs. 42 and 44. **46.** pl. XVI, figs. 13, 16, 17, 23, 25. **47.** pl. XVIII, figs. 1, 4. **48.** pl. XIX. **49.** pl. XX. fig. 17. **50.** figs. 19, 20. **51.** pl. XX, fig. 22*a*. **52.** fig. 21. **53.** fig. 22*b*. **54.** fig. 48. **55.** fig. 21. **56.** figs. 35, 36. **57.** compare figs. 40, 41.

58. compare figs. 40, 42. **59.** fig. 50. **60.** fig. 49. **61.** figs. 45, 46. **62.** fig. 47. **63.** fig. 48. **64.** fig. 26. **65.** 1868*c*: pl. XXIV, figs. 2, 3. **66.** figs. 4–6. **67.** figs. 9–10. **68.** fig. 2. **69.** pl. XXVI, figs. 160, 161. **70.** pl. XXIV, figs. 18, 19. **71.** figs. 25–29. **72.** pl. XXV, fig. 74. **73.** pl. XXIV, figs. 31, 32, 38, 40, 45, 47–49, 62. **74.** figs. 30, 32, 58. **75.** figs. 35, 36, 38, 40, 49–52. **76.** figs. 30, 33, 36, 39, 40. **77.** figs. 43, 44, 47, 48, 50, 51, 56, 58, 59. **78.** figs. 30–36. **79.** figs. 43, 44, 46. **80.** figs. 47, 48, 50, 52. **81.** fig. 42. **82.** figs. 47–49. **83.** figs. 46, 52. **84.** compare figs. 51, 54. **85.** figs. 55, 57. **86.** pl. XXIV, figs. 58–60, 66, 67. **87.** figs. 58, 59. **88.** figs. 60–65. **89.** figs. 68, 69. **90.** fig. 69a. **91.** figs. 70–73; pl. XXV, figs. 74–77. **92.** pl. XXV, fig. 78. **93.** fig. 79. **94.** fig. 80. **95.** fig. 83. **96.** figs. 84, 85. **97.** figs. 86, 87. **98.** fig. 81. **99.** fig. 87. **100.** figs. 88–91. **101.** pl. XXVI, figs. 110, 111, 123–126, 137, 138, 140. **102.** figs. 93–105, 156. **103.** fig. 131. **104.** fig. 152. **105.** figs. 112–114, 117, 119, 123, 126–128, 136, 148–151, 156, 157. **106.** fig. 98. **107.** figs. 109–111. **108.** figs. 98, 109. **109.** figs. 109, 110. **110.** figs. 109, 111. **111.** fig. 111. **112.** figs. 110, 111. **113.** fig. 109. **114.** fig. 110. **115.** fig. 111. **116.** figs. 93–105. **117.** figs. 106–123. **118.** figs. 106, 107, 112, 113, 117, 118, 120, 123. **119.** figs. 124–146. **120.** figs. 124–126, 140. **121.** figs. 132, 134. **122.** fig. 124. **123.** fig. 128. **124.** fig. 130. **125.** figs. 136, 137. **126.** fig. 146. **127.** figs. 147–154. **128.** figs. 155–159. **129.** figs. 149, 156–159. **130.** figs. 150, 151. **131.** pl. XXVII, figs. 160, 164*a*. **132.** figs. 160, 164*b*, 164*c*, 165, 166, 173–176. **133.** figs. 165–167, 169, 174. **134.** figs. 160, 161. **135.** figs. 162, 163, ?. **136.** fig. 167. **137.** pl. XXVIII, figs. 183–196. **138.** fig. 183. **139.** figs. 183–188, 195, 196. **140.** fig. 197. **141.** figs. 202, 206. **142.** figs. 202, 207. **143.** figs. 202, 204, 205, 207–210. **144.** fig. 205. **145.** figs. 211–216. **146.** figs. 218–221. **147.** fig. 223. **148.** fig. 228. **149.** figs. 222, 224, 225, 231. **150.** fig. 229.

REFERENCES CITED

ANONYMOUS. 1905. "Fredrik Adam Smitt." *Kungl. Svenska Vetenskaps-Akademiens Arsbok för år 1903.* 1905: pp. 225–240.

ANNOSOCIA, E. 1968. "Problems of Methodology in Studying and Describing Bryozoa." *Atti della Società Italiana di Scienze Naturali e del Museo Civico di Storia Naturale di Milano* 108: pp. 237–240.

BANTA, W. C. 1968. "The Body Wall of Cheilostome Bryozoa. I. The Ectocyst of *Watersipora nigra* (Canu and Bassler)." *Journal of Morphology* 125: pp. 497–508.

—— 1969. "The Body Wall of Cheilostome Bryozoa. II. Interzoidal Communication Organs." *Journal of Morphology* 129: pp. 149–170.

—— 1970. "The Body Wall of Cheilostome Bryozoa. III. The Frontal Wall of *Watersipora arcuata* Banta, with a Revision of the Cryptocystidea." *Journal of Morphology* 131: pp. 37–56.

BOARDMAN, R. S., and A. H. CHEETHAM. 1969. "Skeletal Growth, Intracolony Variation, and Evolution in Bryozoa: A Review." *Journal of Paleontology* 43: pp. 205–233.

BORG, F. 1926. "Studies on Recent Cyclostomatous Bryozoa." *Zoologiska Bidrag från Uppsala* 10: pp. 181–507.

BUCK, C. D. 1948. [4th Corrected Impression.] *Comparative Grammar of Greek and Latin* (Chicago, University of Chicago Press).

DANIELSSON, U. 1963–1964. "Darwinismens inträngande i Sverige. I." *Lychnos* 1963–1964: pp. 157–210.

—— 1965–1966. "Darwinismens inträngande i Sverige. II." *Lychnos* 1965–1966: pp. 326–334. [With English summary]

DARWIN, C. 1839. *Journal of Researches into the Geology and Natural History of the Various Countries Visited by H. M. S. Beagle* (London, Henry Colburn).

—— 1872. *The Origin of Species* (6th ed., London, John Murray).

DeBEER, G. 1962. *Embryos and Ancestors.* (Third Edition, with additions and corrections, Oxford, Clarendon Press.)

DU CANGE, CHARLES DU FRESNE SIEUR. 1883–1887. *Glossarium mediae et infimae Latinitatis* (Niort, L. Favre; Editio Nova, 10 v.).

ELIAS, M. K. 1971. "Concept of Common Bud and Related Phenomena in Bryozoa." *University of Kansas Paleontological Contributions, Paper 52.*

GHISELIN, M. T. 1969. *The Triumph of the Darwinian Method* (Berkeley, University of California Press).

GHISELIN, M. T., and L. JAFFE. 1973. "Phylogenetic Classification in Darwin's *Monograph* on the Sub-Class Cirripedia." *Systematic Zoology.* 22: pp. 132–140.

GILDERSLEEVE, B. L. 1895. [1963 reprint used.] *Latin Grammar* (3rd ed., London, Macmillan).

GRADENWITZ, O. 1904. *Laterculi vocum Latinarum* (Leipzig, S. Hirzel).

HINCKS, T. 1880. *A History of the British Marine Polyzoa* (London, John Van Voorst, 2 v.).

—— 1890. "Critical Notes on the Polyzoa; XII." *The Annals and Magazine of Natural History* [ser. 6] 5: pp. 83–103.

HOFMANN, J. B. 1965. *Lateinische Syntax und Stilistik.* Revised by A. Szantyr (Munich, Verlag C. H. Beck).

HYMAN, L. H. 1940. *The Invertebrates: Protozoa through Ctenophora* (New York and London, McGraw-Hill Book Co.).

NEUE, F. 1892–1905. *Formenlehre der lateinischen Sprache.* 3rd Edition, greatly enlarged by C. Wagener. 1 (Leipzig, O. R. Reisland, 1902); 2 (Berlin, S. Calvary und Co., 1892); 3 (Berlin, S. Calvary und Co., 1897); 4 (Leipzig, O. R. Reisland, 1905).

RYLAND, J. S. 1967. "Polyzoa." *Oceanography and Marine Biology Annual Review* 5: pp. 343–369.

—— 1970. *Bryozoans* (London, Hutchinson University Library).

SENECA. 1969. *Letters from a Stoic; Epistulae morales ad Lucilium.* [A.D. 62–64]. Selected and translated with an introduction by R. Campbell (Baltimore, Penguin Books).

SCHOPF, T. J. M., and D. F. TRAVIS. 1968. "Skeletal Wall Structure of a Calcified Bryozoan (Phylum Ectoprocta)." *Biological Bulletin* 135: p. 436.

TAVENER-SMITH, R., and A. WILLIAMS. 1970. "Structure of Compensation Sac in Two Ascophoran Bryozoans." *Proceedings of the Royal Society of London,* B 175: pp. 235–254.

THESAURUS LINGUAE LATINAE. 1904–1953. (Leipzig, Teubner, 9 v. to date).

TRANSACTIONS

OF THE

AMERICAN PHILOSOPHICAL SOCIETY

HELD AT PHILADELPHIA
FOR PROMOTING USEFUL KNOWLEDGE

NEW SERIES—VOLUME 63
1973

THE AMERICAN PHILOSOPHICAL SOCIETY
INDEPENDENCE SQUARE
PHILADELPHIA

1973

CONTENTS OF VOLUME 63